Path Planning of Cooperative Mobile Robots Using Discrete Event Models

Path Planning of Cooperative Mobile Robots Using Discrete Event Models

Cristian Mahulea
University of Zaragoza, Spain

Marius Kloetzer
"Gheorghe Asachi" Technical University of Iasi, Romania

Ramón González
Robonity, Spain

IEEE PRESS
WILEY

Published by John Wiley & Sons, Inc., Hoboken, New Jersey.
Published simultaneously in Canada.

Library of Congress Cataloging-in-Publication Data is available.

Hardback: 9781119486329

Set in 9.5/12.5pt STIXTwoText by SPi Global, Pondicherry, India

Printed in the United States of America

V10016323_121619

To our families

Contents

Foreword

The path planning approach based on Discrete Event System (DES) models is an important – but a challenging – problem. It has been studied for a number of years in the DES community in the field of automated guided vehicle (AGV) systems. In this case, the main issue is to compute collision-free paths for an AGV from a starting configuration to a goal one. For multi-agent systems, an additional problem is to avoid deadlocks that could result if waiting modes are introduced. The main solutions obtained are based on the results of deadlock prevention and avoidance from Resource Allocation Systems that allow one to use a DES structure to obtain deadlock-free trajectories of AGVs. Even if the complexity of these approaches is high, researchers have identified some computationally tractable solutions for some particular classes.

This book is a step forward from the classical AGV navigation problem, by assuming high-level specifications for a team of cooperative robots. Therefore, the problem is not to reach only some goal configurations, but to accomplish a more complicated task combining logic and temporal operators on some regions of the environment, called regions of interest. This problem appeared some years ago in the formal method community, where so-called symbolic approaches are used to solve it. Mainly, its solution consists in obtaining a model for a robot team and a model for specifications, combining them in a smart way and then computing robot trajectories. Initially, the model of the team is obtained by using transition systems that, in the case of teams with more than 2 or 3 robots, suffer from a so-called state space explosion problem, i.e. the number of discrete states grows exponentially with the number of robots, making this approach impractical.

In order to avoid this state space explosion problem, two alternative approaches have been proposed. One approach consists in dividing a high-level specification into (local) subformulas that can be executed in parallel, and then, for each local formula, a special model is constructed allowing distribution of the fulfillment of the task among robots. Another approach, which is presented in this book, tackles the state explosion problem by using Petri net models and optimization techniques, thus avoiding a formula from being divided and, hence, allowing more general specifications than language subclasses.

This book presents an open-source Matlab toolbox called Robot Motion Toolbox (RMTool), which can be freely downloaded and used to check all the presented approaches. Chapter 2 introduces this toolbox and focuses on its usage in introductory courses of robotics. All the examples in the following chapters are illustrated by using this toolbox. A reader can repeat them easily.

DES models can be mainly constructed by using two approaches: cell decomposition of an environment and sample-based methods. This book is focused on the former as presented in Chapter 3. It allows one to obtain a DES model for a team of robots. The main advantage of this approach over the sample-based one is the fact that it always returns a solution in finite time as long as this solution exists. The formal description of the necessary DES models is given in Chapter 4, together with the high-level specifications to be used: Linear Temporal Logic,and Boolean-based formulas.

For completeness, this book also presents the transition system models and methods. Chapter 5 presents some results related to this type of model that basically consists in computing the synchronous product of different transition systems in order to obtain the model of a team of robots. Some new techniques for computing trajectories are also presented, as for example one based on Model Predictive Control. Furthermore, it describes a method to avoid collisions by introducing initial delays to some trajectories.

One of the main novelties of the book is the approaches described in Chapter 6 based on Petri net models, which are defined in Chapter 4. These approaches try to avoid the construction of the synchronous product of automata. However, an additional problem appears and is related to obtaining the sequence of firing transitions. The optimization problems used for path planning return firing vectors, but, in the current case, firing sequences are necessary to obtain robot moving sequences. In general, this is a very complicated problem in the Petri net literature. However, based on the particular structure of a Petri net, an algorithm is provided to compute the firing sequence in polynomial time. Additionally, Chapter 6 presents some particular problems related to the collections of some tasks and discusses further approaches for deadlock prevention.

This book opens a number of future research directions and can be of particular interest in two different communities. First, for the Discrete Event System community, it opens an interesting application area of being possible to apply the results from this community to a particular application. Second, for the robotic community, it can use the theoretical results from a different community to solve particular problems. Hence, this book should certainly become a very important addition to the IEEE Press–Wiley Book Series on Systems Science and Engineering.

MengChu Zhou, PhD and Distinguished Professor
Fellow of IEEE, AAAS, IFAC, and CAA
Founding Editor, IEEE Press–Wiley Book Series on Systems Science and Engineering
Email: zhou@njit.edu
Website: http://web.njit.edu/ zhou

Preface

Mobile robotics comprises a successful field in the world today. It is not strange to see little mobile robots cleaning our house, or big robots moving goods in factories, harbors or airports, or even adventurous mobile robots exploring other worlds.

One common issue to all those robots deals with the problem of generating feasible paths or routes between a given starting position and a goal or target position while avoiding (static) obstacles. This problem is addressed within the area of path planning. Due to the importance of this problem in robot navigation, path planning has received considerable attention and numerous strategies have been proposed.

When a group of robots work within the same environment and cooperate in order to accomplish a high-level task given as a high-level specification, standard path planning algorithms employed by the robotics community, based on potential functions or road maps, may lead to wrong or even unfeasible results.

This book formulates the problem of path planning of cooperative mobile robots by using the paradigm of discrete-event systems. First, a high-level specification is expressed in terms of a Boolean or Linear Temporal Logic (LTL). The environment is then divided into discrete regions of a chosen geometrical shape by using cell decomposition. This book compares the performance of several cell decomposition algorithms in terms of several metrics. This decomposition can be used to define a discrete event system (DES) modeling the movement capabilities of the robot or of the team by using Transition System or Petri Net models. The obtained DES is next combined with the model of the high-level specification to be accomplished by the group of robots. Finally, the resulting model is used to compute the trajectories via a graph search algorithm or solving optimization problems.

This book contributes an interactive software tool that the intended user can exploit in order to simulate and test all the strategies introduced and formulated in the book. This software tool, called RMTool (Robot Motion Toolbox), is freely available online and can be run in Matlab. It can be used for teaching mobile robotics in introductory courses, as the user can interact with the tool by using a Graphical User Interface (GUI), without requiring previous knowledge of Matlab.

This book is primarily aimed at undergraduate and graduate students and college and university professors in the areas of robotics, artificial intelligence, systems modeling, and autonomous control. The topics addressed in this book can also be welcomed by researchers, PhD students, and postgraduate students with a focus on robot motion planning, centralized robot planning solutions for teams of robots, and interactive teaching tools to be used in engineering courses. The contents of this book and the accompanying software tool can be employed by students and professors at the high-school level with a previous background in mathematics and engineering.

Zaragoza, Spain *Cristian Mahulea*
Iasi, Romania *Marius Kloetzer*
Almeria, Spain *Ramón González*

Acknowledgments

The authors would like to express their sincere appreciation to all their collaborators, especially to the following professors and researchers (in alphabetic order) who co-authored some works that further led to the results included in this book: Calin Belta, Adrian Burlacu, Yushan Chen, José-Manuel Colom, Xu Chu Ding, Narcis Ghita, Karl Iagnemma, Doru Panescu, Luis Parrilla, Octavian Pastravanu, Manuel Silva, Emanuele Vitolo, and Xu Wang. Many thanks to Professor MengChu Zhou for the invitation to write this book for the IEEE Press–Wiley Book Series on Systems Science and Engineering.

Furthermore, our thanks go to the institutions that offered support for performing the research on which this book is based: Aragón Institute on Engineering Research (I3A) and Department of Computer Science and Systems Engineering, University of Zaragoza, Spain; Faculty of Automatic Control and Computer Engineering, Technical University of Iasi, Romania; Robonity: innovation driven startup, Spain; Center for Information and Systems Engineering, Boston University, USA; Massachusetts Institute of Technology, USA.

The authors also acknowledge the financial support of the following grants from the last few years. In Spain: MINECO-FEDER DPI2014-57252-R project, University of Zaragoza UZ2018-TEC-06 and JIUZ-2018-TEC-10 projects and CEI Iberus Mobility Grants 2014 funded by the Ministry of Education of Spain within the Campus of Excellence International Program; in Romania: CNCS-UEFISCDI project PN-III-P1-1.1-TE-2016-0737, CNCSIS-UEFISCSU project PN-IIRU-PD-333/2010; in China: NSFC Grant No. 6155011023.

We express our thanks to those who read the initial version of this book and formulated useful suggestions, namely the anonymous reviewers and our colleagues Eduardo Montijano and Sofia Hustiu. Last but not least, the authors are most grateful to their families for all their love, encouragement, and support.

Acronyms

AGV	Automated Guided Vehicle
CNF	Conjunctive Normal Form
CPU	Central Processing Unit
CTL	Computation Tree Logic
DNF	Disjunctive Normal Form
FSA	Finite State Automata
GUI	Graphical User Interface
GVD	Generalized Voronoi Diagram
ICR	Instantaneous Centre of Rotation
LTL	Linear Temporal Logic
MILP	Mixed-Integer Linear Programming
MPC	Model Predictive Control
ODE	Open Dynamics Engine
PI	Proportional Integral
PID	Proportional Integral Derivative
PN	Petri Net
PRM	Probabilistic Road Map
RARMPN	Resource Allocation Robot Motion Petri Net
RAS	Resource Allocation System
RMPN	Robot Motion Petri Net
RMTool	Robot Motion Toolbox
RRT	Rapidly exploring Random Tree
V-Graph	Visibility Graph

1

Introduction

1.1 Historical perspective of mobile robotics

Since its first application in the 1940s, robot arms or manipulators have demonstrated a great success in the world of industrial manufacturing. These robot arms can perform repetitive tasks such as spot welding, painting, machine loading and unloading, electronic assembly, packaging, and palletizing, among other activities. However, industrial robots lack of one fundamental property: mobility. The fixed-base manipulator has a limited range of motion that depends on where it is bolted down. The ability to move is what makes a mobile robot travel freely throughout a given environment. However, this mobility advantage can also be its doom if the robot does not account for a reliable navigation strategy.

One navigation approach is to just react to what is sensed; this is called reactive navigation [7, 186, 193]. For example, the robotic tortoise Elsie, built in the 1940s by the Edison-Swan Electric Company, reacted to her environment and could seek out a light source without having any explicit plan or knowledge of the position for the light, see Figure 1.1a. This reactive navigation strategy is exploited today by Automated Guided Vehicles (AGVs) in many factories [6, 144]. For example, Amazon uses AGVs in more than ten of its warehouses located in the US. These robots were developed by the company Kiva, later acquired by Amazon in 2012 and becoming AmazonRobotics (https://www.amazonrobotics.com).

Reactive systems can be fast and simple when sensing is connected directly to action, that is, there is no need for resources to hold and maintain a representation of the world nor any capability to reason about that representation [41]. However, such reactive navigation requires a fixed infrastructure where the robot is going to move, for example, a painted line on the floor, a buried cable that emits radio-frequency signals, or wall-mounted bar codes. The second major drawback of this approach is that it limits the mobility of the robot to those areas where the guidance system is located or installed; this explains why AGV are usually applied in factories.

Path Planning of Cooperative Mobile Robots Using Discrete Event Models, First Edition.
Cristian Mahulea, Marius Kloetzer, and Ramón González.

(a) ELSIE robot (1949).
Source: Cyberneticzoo

(b) Amazon-Kiva robot (2008).
Source: Amazon-Robotics

Figure 1.1 Mobile robots and reactive navigation. Example of mobile robots based on reactive navigation strategies.

With the explosion of digital technology in the 1970s, a group of engineers working at the Stanford Research Institute (SRI) developed the first mobile robot to be operated using autonomous reasoning [159]. The robot Shakey was capable of 3D perception and created a map of its environment and then reasoned about the map to plan a path to its destination, see Figure 1.2a. An optical rangefinder and a vidicon television camera with controllable focus and iris were mounted on a tilt platform for sensing. Offboard communication was provided via two radio channels: one for video and the other providing command and control. This ability to make maps and reason about them made Shakey capable of performing more complex tasks than former robots based on reactive navigation strategies. After Shakey, the next big step in the field of mobile robotics came from the Robotics Institute at Carnegie Mellon University in the 1980s [28]. The Terregator robot (Terrestrial Navigator) was designed to operate autonomously in outdoor scenarios, see Figure 1.2b. It was a six-wheeled skid-steer robot utilizing compliant tires for suspension. Terregator subsystems included locomotion, power, computation, controls, wireless telemetry (serial links and two channels of UHF video), orientation sensors, and navigation payloads. The robot's onboard control system consisted of a central processing unit (CPU) linked to motor controllers. This CPU calculated the robot's position and orientation from a gyroscope, wheel encoders, and inclinometers (dead-reckoning). Additionally, this system was also responsible for guiding the robot to a commanded destination provided by an offboard supervisor (remote command station). This supervisor made use of the images coming

(a) Shakey robot (1969). Source: http://www.ai.sri.com/shakey

(b) Terregator robot (1985) [29]

(c) Mercedes autonomous vehicle (1988) [53]

Figure 1.2 Autonomous mobile robots. Example of pioneering autonomous mobile robots.

from the video camera mounted on the vehicle. Terregator was even used for mapping a portion of a mine thanks to its path planning and mapping capabilities [28]. Another big milestone in the early ages of mobile robotics came from the University of Munich in Germany. Several vehicles developed by Prof. Ernst Dickmanns, e.g. Daimler-Benz VITA-2 and UniBwM VaMP, drove autonomously on European highways at speeds up to 130 km/h, see Figure 1.2c. This astonishing leap was achieved by using a visual feedback control system that tracked the road boundaries. This advanced control system ran in a multiprocessor image processing system using contour correlation and curvature models together with the laws of perspective projection [52]. The vehicle's position was estimated relative to the driving lane and road curvature by means of a Kalman filter. In 1994, the VaMP driverless car designed by Prof. Dickmanns drove 1600 kilometers, of which 95% were driven autonomously [51].

These pioneering mobile robots opened the door to the significant achievements in the field of mobile robotics during the last twenty years. Today, mobile robots have explored other worlds such as Mars or the Moon, e.g. MER rovers and MSL rover [87, 194]; mobile robots have worked at Antarctica seeking meteorites, e.g. Nomad robot [5]; robots are able to clean our houses, e.g. iRobot Roomba; to perform harvesting activities in agriculture, e.g. ASI autonomous tractor; and they are also present in our schools, like the SoftBank Robotics NAO humanoid robot.

1.2 Path planning. Definition and historical background

A fundamental task for any mobile robot is its ability to plan collision-free trajectories from a start to a goal position or visiting a series of positions, i.e. regions of interest, among a collection of (static) obstacles. This is the robot's cognitive level. Cognition generally represents the purposeful decision-making and execution that a system utilizes to achieve its highest-order goals. Given a map and a goal position (or a list of high-level goals), path planning involves identifying a trajectory that will cause the robot to reach the goal position (a 2D point) or pose (a 2D point and an orientation) when executed [35, 135, 191]. It bears mentioning that position refers to the longitudinal and lateral coordinates in a Cartesian frame. Pose also considers the orientation.

Path planning comprises the highest level within a typical navigation architecture of a mobile robot. As observed in Figure 1.3, there are four big modules or layers. The second layer in this navigation architecture includes the motion controller (or trajectory following strategy), which is responsible for generating the control actions in such a way that the robot follows as accurately as possible the reference trajectory provided by the path planning layer. These control actions will be determined according to the error between the current robot's position and the next desired waypoint. The third layer deals with the low-level PID controllers that ensure that the control actions generated by the motion controller are reached by the robot actuators, i.e. wheel motors. As explained before, it is crucial for the motion controller to know the current position of the robot; this is calculated by the last layer in the robot's navigation architecture: the localization layer. Here, a set of sensors are employed to get the robot's position as accurately as possible. Another significant observation about this traditional navigation architecture is that the execution time or the sampling time of each layer is different. For example,

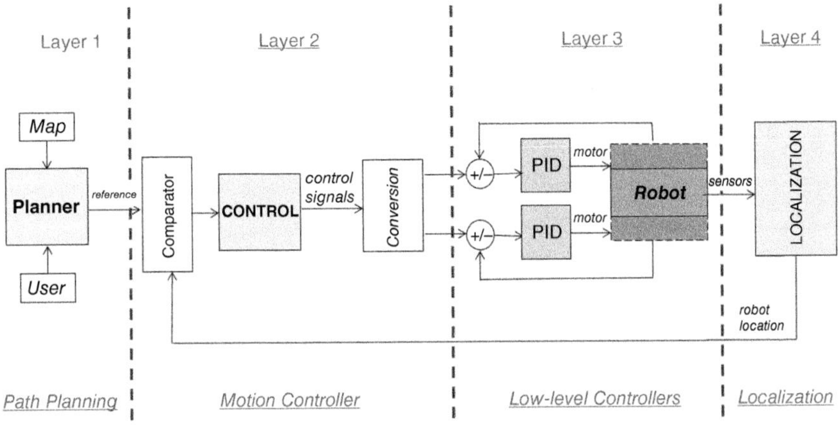

Figure 1.3 Navigation architecture of a mobile robot. Example of a navigation architecture of a mobile robot (University of Almeria's Fitorobot robot) [78].

the time to get a response from the path planning layer is on the order of minutes or seconds, the time for the motion control layer is on the order of seconds or hundreds of milliseconds, and the fastest layer comprises the low-level controllers that run on the order of milliseconds.

The history of robot path planning began with the early computer-controlled robots, i.e. SRI Shakey robot. One of the pioneering references in this field is the book "Robot Motion Planning" authored by Jean-Claude Latombe in 1980 [132][1]. At that time, the most advanced planners were barely able to compute collision-free paths for objects moving in planar workspaces. In the 1990s, researchers working in this field faced another challenge: planning motions in the presence of kinematic constraints like non-holonomic robot systems[2], e.g. a car that cannot rotate around its axis without also changing its position [133]. At the beginning of the twenty-first century, robot planners could be applied to robots with many degrees of freedom in complex environments, coordinate multiple robots, and handling dynamic environments. In 2005 and 2006 other key references appeared in the field of mobile robotics: "Principles of Robot Motion: Theory, Algorithms, and Implementations" by Choset et al. [35] and "Planning Algorithms" by Stephen LaValle [135]. These books opened the door to multiple milestones in this field, as roadmaps, cell decomposition methods, and sampling-based planning algorithms.

1 Motion planning in a broad sense involves the generation of inputs to a nonlinear dynamical system that drives it from an initial state to a specified goal state [135]. In this sense, path planning can be considered as a subfield within motion planning.

2 Non-holonomic systems are, roughly speaking, mechanical systems with constraints on their velocity that are not derivable from position constraints [17].

Traditional planning algorithms, sometimes called combinatorial or exact algorithms, build discrete representations of a given environment, i.e. a map, without losing any information. Some of them are complete, which means that they must find a solution if one exists; otherwise, they report failure [135]. On the contrary, sampling-based solutions sparsely sample the world map and conduct discrete searches that utilize these samples. This paradigm sacrifices completeness with the benefit of a faster computation [135].

As Figure 1.4 shows, there are two big areas in the field of path planning in mobile robotics: combinatorial or exact algorithms and sampling-based planning. This book focuses on the first category, but the second category is also described for completeness purposes. Notice that the area of reactive navigation has also been included here for completeness purposes. Recall that the main difference between path planning approaches and reactive navigation is that in the first case a map of the environment is needed.

The first category, exact algorithms, can be divided into two areas: road map planning and cell decomposition. The road map planning approaches capture the connectivity of the robot's free space in a network of 2D curves or lines, called road maps. Once a road map is constructed, it is used as a network of road segments. At this point, path planning is reduced to connecting the initial and goal position of the robot to the road network and then searching for a series of roads from the initial robot position to its goal position [191]. To this category belong two well-known methods in the field of mobile robotics: visibility graph and Voronoi diagram.

The visibility graph (or V-graph), formally described by Lozano-Perez and Wesley in the 1970s [142], represents a complete and easy to implement algorithm. This algorithm is based on constructing an undirected graph where edges come as close as possible to obstacles, then resulting in minimum-length paths. An important aspect is that obstacles can be inflated in order to avoid an incident where the robot could pass by too close to them, which could lead to collisions [58]. The main drawbacks of this algorithm are that it can demand a high computation time for getting a trajectory in environments with complicated obstacles, and some points of the path are too close to obstacles if inflation is not used. The fast dynamic visibility graph (DVG) approach proposed in [100] represents an efficient implementation of the traditional V-graph. The V-graph has been largely used by the robotics community from Shakey in the 1970s to recent publications like [39], where the V-graph algorithm is used to find the obstacle-free path after processing digital images acquired by a camera onboard a mobile robot.

The Voronoi diagram was proposed by G.F. Voronoi at the end of the nineteenth century and since then it has been widely used in many areas such as: mathematics, physics, astronomy, geographical information systems, computer graphics,

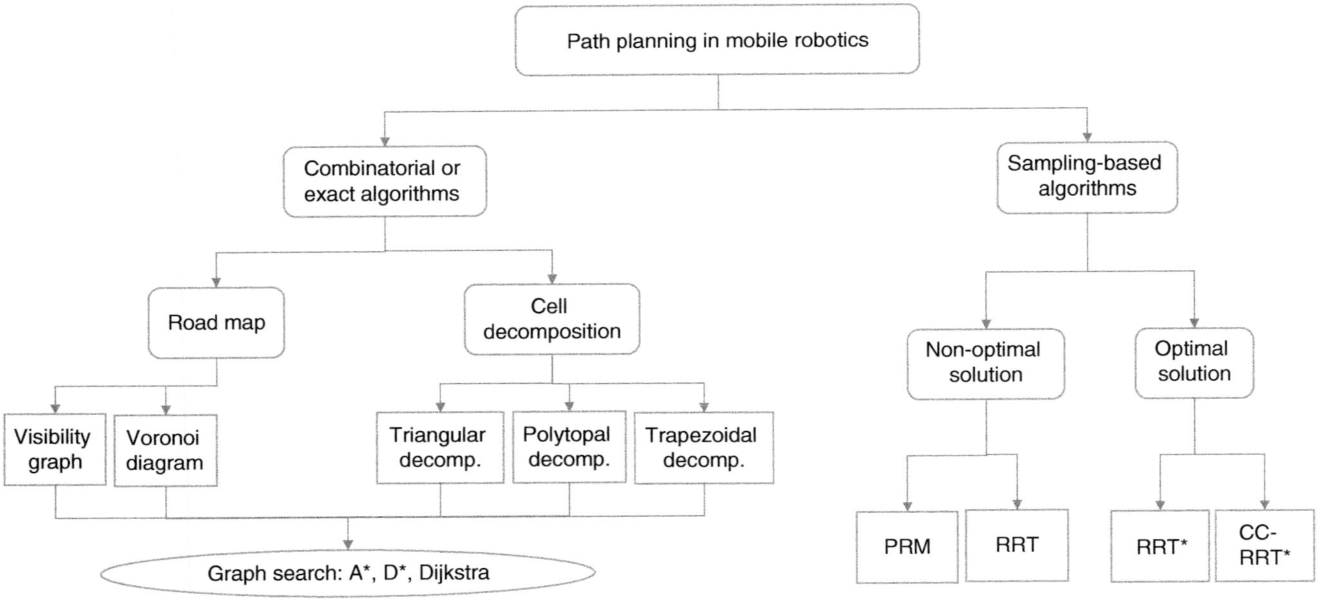

Figure 1.4 Path planning methods. Main categories for path planning in mobile robotics and examples of some well-known methods for each of them.

image processing, and robotics. In contrast to the V-graph approach, a Voronoi diagram tends to maximize the distance between the robot and obstacles in the map, that is, roads stay as far away as possible from obstacles. Therefore, there is no need to inflate the obstacles. As in the V-graph after the set of roads are determined, a graph search algorithm is applied to estimate the best route between the desired points. An application of the Voronoi diagram in mobile robotics appears in [79]. Here, this algorithm is used by an agricultural robot to move in a greenhouse. There is also a broad field of research combining the advantages of the visibility graph (shortest routes) with those of the Voronoi diagram (maximum distance from the obstacles); see, for example, [151, 211].

The second big category comprising exact planning algorithms is cell decomposition. This strategy is based on partitioning the free configuration space into a finite set of regions that can be safely traversed by a robot. Cell decompositions are often employed in high-level planning approaches where the robot may visit some regions based on logic or temporal requirements (this feature is extended in Section 1.4 and in subsequent chapters).

It bears mentioning that the exact methods further need graph search algorithms in order to finally obtain the optimal route connecting a series of waypoints. For this aspect, three graph search algorithms that return a minimum cost path are usually employed, namely Dijkstra [53], A* [140, 160, 221], and D* [195]. Dijkstra and A* algorithms are great choices for static graphs with positive known arc weights, with A* additionally requiring a heuristic cost function. The D* algorithm, proposed by Anthony Stentz in 1995 [195], constitutes an extension of the classic A* algorithm for graphs, introduced by Nils J. Nilsson in the 1970s [160]. D* considers the 2D map as a cost map where each weight represents the cost of traversing each cell in the horizontal or vertical direction. For cells corresponding to obstacles we can give a huge cost, so D* finds the path minimizing the total cost of travel. This cost may deal with various aspects such as time to drive across a cell, roughness of the terrain, etc. The key feature of D* is that it supports incremental replanning, that is, the algorithm can replan the initial route while the robot discovers that the world is different from the initial plan. The incremental replanning has a lower computational cost than completely replanning, as would be required by A* [41]. The algorithm D* accounts for many successful applications in mobile robotics. For example, in [84], the D* algorithm is used for generating the optimal route for ground autonomous vehicles. More specifically, the cost function associated with the D* algorithm is configured in such a way that the route connecting two points minimizes the uncertainty associated with the elevation of the 3D points in the maps. This route also avoids points where the robot may experience a high risk of entrapment. After the successful introduction of the D* algorithm, there is a recent and broad body of research dealing with extending the features of this algorithm. Some of those recent algorithms are: D* Lite [125], PHI* [57], and E* [169].

Together with combinatorial or exact planning algorithms, another broad body of research in the field of path planning nowadays is related to sampling-based planning methods. The first algorithm, called Probabilistic Roadmap (PRM), was proposed by Lydia E. Kavraki and Jean-Claude Latombe in the 1990s [106]. The advantage of PRM is that relatively few points need to be tested to ascertain that the points and the paths between them are obstacle free [41]. The efficacy of several variations of the PRM algorithm is discussed in [75].

A major drawback of PRM is that it assumes that the robot is a point with omnidirectional capabilities. The Rapidly exploring Random Tree (RRT) algorithm takes into account the model of the robot, e.g. differential-drive motion [134]. However, the main drawback of RRT is that it does not lead to an optimal path. This aspect is overcome by a variant of RRT called RRT*; this algorithm does guarantee the optimality and can find the optimal trajectory when applied to complex non-holonomic systems [2, 104, 138]. In recent literature, there are numerous RRT-based strategies trying to ensure optimality despite uncertainty in the motion of the robot [136, 143].

1.3 Motion control. Definition and historical background

The key goal of a mobile robot is to follow the route generated by the path planner, and this goal is responsible for the motion controller. More specifically, robot control deals with the problem of determining the forces (or velocities) that must be developed by the robotic actuators in order for the robot to go to a desired position, track a desired trajectory, and, in general, perform some tasks with desired performance requirements [202].

The control problem can be classified depending on how the reference path is defined: a single target point (x, y), a set of waypoints $(x(t), y(t))$, or a set of poses $(x(t), y(t), \theta(t))$, where $t \in \mathbb{R}$ is the time. The time in the previous variables represents a situation where those variables may change along the robot operation. In this sense, the control strategy can be understood as the problem of moving to a point, path following, and trajectory tracking, see Figure 1.5. Notice that, in the trajectory tracking problem the robot orientation must also be controlled, unlike in the other two cases where the final robot orientation depends on the starting position.

Essentially the control problem must ensure

$$\lim_{t \to \infty} e(t) \cong 0, \tag{1.1}$$

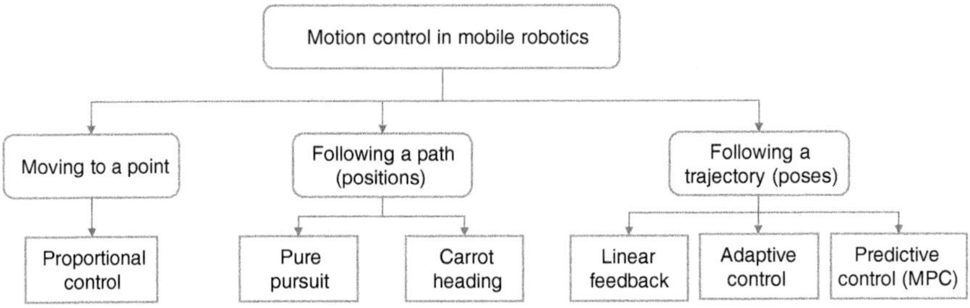

Figure 1.5 Motion control methods. Main categories for motion control in mobile robotics and an example of some well-known methods for each of them.

where e is the error between the actual position of the robot and the desired target position. Notice that Eq. (1.1) defines a quasi-zero error because in some situations, for instance considering uncertainty, an exact error equal to zero cannot be achieved [81]. The control problem associated with a mobile robot can then be defined as a feedback control system. The idea is that the controller senses the position/pose of the robot, compares it against the desired reference, computes corrective actions based on a model of the robot and actuates the robot to effect the desired change. As highlighted in [9], the key issues in designing control logic are ensuring that the dynamics of the closed-loop system are stable (bounded disturbances give bounded errors) and that they have additional desired behavior (good disturbance attenuation and fast responsiveness to changes in the operating point, among others).

It is important to remark that mobile robotics comprises a challenging field from a control standpoint as there are some phenomena that influence robot's controllability, such as hard constraints fulfillment (e.g. physical limitations of actuators, narrow workspaces), and uncertainties (e.g. unmodelled dynamics, simplified models, noisy measurements). For that reason, in the past few years, many research efforts have been devoted to the application of different control strategies.

One of the first path following approaches was proposed by R. Craig Coulter in the early 1980s for the Terregator robot [45]. The name "pure pursuit" comes from the analogy for the way humans drive. We tend to look some distance in front of the car and head toward that spot. This lookahead distance changes as we drive to reflect the twist of the road and vision occlusions. In the pure pursuit algorithm, this lookahead distance plays a key role between accuracy following the desired route and aggressiveness of the control actions [79]. A similar algorithm to pure pursuit is called "carrot heading". Here, the algorithm determines the next point to be followed by calculating the intersection of a circle centered on the rover frame origin with the desired path [94]. These two well-known approaches belong to the category of path followers, that is, the robot follows a desired path with no concern about the orientation.

The second major problem dealing with mobile robot control is the trajectory tracking problem. In this case, the robot must follow a "virtual robot" (position and orientation) at each sampling time [81, 163]. This problem has benefited from the application of advanced controllers that appeared in the field of state feedback control [9, 24, 202]. For example, in [24, 105], a linear feedback control strategy is used for controlling a non-holonomic mobile robot where stability is ensured by tuning the feedback gains of the control strategy according to a Lyapunov function. This controller was extended in [80] for compensating for longitudinal slip in mobile robots operating in off-road conditions.

The problem of trajectory tracking in mobile robotics has also benefited from another broad body of research: Model Predictive Control (MPC). MPC is based on generating the control input to be applied to the robot by solving an optimization problem considering the robot constraints and the desired reference to be tracked [156, 177, 208]. For instance, in [137], the authors use an MPC controller to enable both anticipation of approaching curvature and to compensate from lateral slip phenomena for path tracking control of an agricultural vehicle. In [108], an MPC is applied to the trajectory tracking problem. The control law is analytically derived, which permits its application to a physical mobile robot. In order to avoid vehicle slip, velocity and acceleration are bounded. The work [77] presents a predictive strategy that permits the robot to avoid unexpected static obstacles in the robot environment. In this case, a neural network was trained to be able to run the MPC-based controller in real-time. A Smith-predictor-based generalized predictive controller is discussed in [161]. This control strategy permits dealing with dead-time uncertainties related to a mobile robot control motion. Generally, the main issue of robust MPC strategies, which sometimes prevent its physical application, is related to the high computation burden [176]. Recently, an efficient theoretical concept, called a "tube-based MPC", has been applied to robustify MPC and has been applied to mobile robots operating in off-road conditions [11, 80, 81].

1.4 Motivation for expressive tasks

As previously explained, path planning may be understood as a strategic competence that deals with the long-term goal of a robot (or a group of robots), that is, depending on the mission to be achieved by the robot different routes or regions must be visited or avoided based on logic or temporal requirements [13, 14, 66, 88, 118]. As [13, 14] explain, this way to define the path planning problem can be solved by following a hierarchical, three-level process (see Figure 1.6). At the top of this hierarchy is the specification level. Here is exactly where strategies based

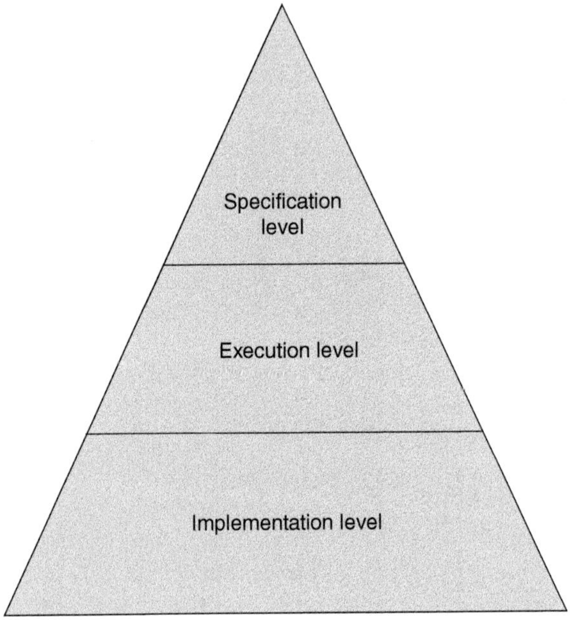

Figure 1.6 Path planning levels. Path planning levels of standard navigation problem.

on cell decomposition play a key role. Recall that cell decomposition is based on partitioning the free configuration space into a finite set of regions that can be safely traversed by the robot. These set of regions can be represented as a graph. After this, the execution level is responsible for generating an optimal path, for example, by minimizing the expected traveled distance and/or accounting for the proximity to obstacles. Finally, the implementation level ensures that the reference trajectory traversing the sequence of cells given by the path previously obtained is followed by the actual robot. The main limitation of this way to formulate the path planning problem is that specifications like "reach cell A, and then cell B, and stay there for all future times" cannot be properly addressed. This is what motivates the use of symbolic approaches for the specification of high-level path planning missions [13, 66].

One possible necessity of using high-level specifications instead of standard navigation problem (reaching a final configuration) is in the factories of the future, where workers and robots will cooperate in fulfilling complex tasks, in order to obtain as soon as possible the products requested by the consumers [55, 179, 189]. The robots should automatically adapt and optimize the usage of these shared frameworks [50, 197], and one of the important aspects is to automatically compute collision-free trajectories. Since the number of mobile robots could be high, it is important to have computationally attractive techniques to plan trajectories for teams of robots.

There are several formalisms for specifying the high-level path planning goals explained previously, from which we focus on two: Linear-time Temporal Logic (LTL) formulas [10] and Boolean-based formulas [147]. These approaches and their applications will be largely explained in subsequent chapters, as only a brief description of them is given here. Other formalisms, such as Computation Tree Logic (CTL) [10] and generalized reactivity (1) formula (GR(1)) [170], are also employable to capture similar specifications.

Linear-time temporal logic is a general-purpose mathematical language for describing linear-time problems, which was first pointed out by Amir Pnueli in 1977 [171]. In the specific context of mobile robotics, the set of mathematical operators defined by LTL can be used for encoding rules about the sequence of paths that a robot (or a team of robots) should follow. The success of LTL is that high-level specifications defined in a natural human level can be easily translated into LTL statements. Although LTL formulation has demonstrated a significant contribution to the field of mobile robotics [13, 66, 111], it generally demands a high computation load.

Figure 1.7 shows an example involving a path planning problem that can be solved by using the LTL formulation. Imagine the user wants to use the robot for performing a surveillance activity. As a result, he/she defines the following high-level mission: "the robot must reach A and then B, once it reaches B, the robot must move from B to A and back constantly". The first step would be to run a cell-decomposition-based algorithm in order to partition the free configuration space into a finite set of regions. After that, the LTL formula and a suitable approach would find the sequence of steps to go from the current position of the

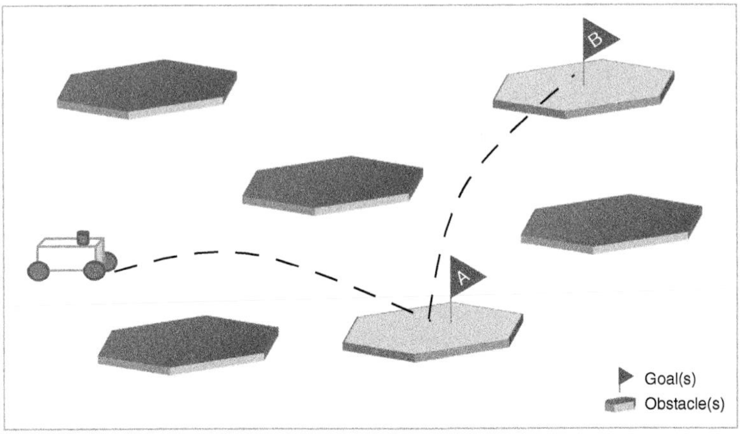

Figure 1.7 Expressive tasks. Meaning of expressive tasks.

robot to A and then to B. Finally, the robot would move from B to A and back constantly. Should the mission required only "the robot must reach A and then B", a Boolean-based formalism for capturing it and a corresponding method based on Petri nets could be used, as it will be detailed in subsequent chapters. Notice that at the implementation level, the lowest level, the motion controller should ensure that the robot follows as closely as possible the trajectories generated by the path planning execution level.

1.5 Assumptions of this monograph

At this point, two primary assumptions related to the work presented in this monograph are highlighted.

The first assumption is that the mobile robots are considered as omnidirectional points $(x, y) \in \mathbb{R}^2$, where the domain of operation is a continuous Cartesian plane (2D plane). In this sense, the path planning problem can now be stated as moving the robot from its initial position (x_i, y_i) to another position (x_g, y_g) (or list of high-level goals) while avoiding obstacles or while driving over the free space.

The second assumption is that the robot evolves in a known and static 2D environment cluttered with convex polygonal obstacles. This means that the robot should move from one position to another while remaining within C_{free}, where C_{free} is an obstacle-free space for the map where the robot is going to move.

1.6 Outline of this monograph

This book is composed of seven chapters. This current chapter serves as an introduction and gives an historical perspective of two important topics: path planning and motion control in the field of mobile robotics.

Chapter 2 presents a Matlab-based interactive software tool, called Robot Motion Toolbox (RMTool) [82, 165]. With RMTool, the interested reader can test various theoretical concepts introduced throughout this monograph. The software tool offers the user the advantage of instantly seeing the motion planning results in a Graphical User Interface. Therefore, there is no need for previous knowledge in Matlab or in programming the methods and algorithms tackled in this book. However, the available functions can also be used outside RMTool for implementing new algorithms. The chapter includes illustrative examples demonstrating the benefits of the proposed simulator, and at the same time it informally introduces some planning strategies that are to be detailed in the subsequent chapters.

Chapter 3 focuses on cell decompositions, which will be further used for automatically planning mobile robots. The chapter discusses the algorithms for performing four types of decompositions for bidimensional environments [76, 114]. Details on the available Matlab implementation are also given, the mentioned functions being part of RMTool. The chapter includes different qualitative and quantitative comparisons regarding the relative performance of the four cell decompositions.

Chapter 4 defines the discrete event system models that are to be used for abstracting the motion of a single robot or of a team of identical robots in a given environment. Two types of models are constructed by using cell decompositions, namely transition systems and Petri nets. The chapter presents some high-level specifications that can be used for imposing desired robot behaviors. Based on the constructed models and on specifications, the following chapters will present automatic methods for planning the robotic motion.

Chapter 5 focuses on planning based on transition system models. First, algorithms are presented for classical planning of a robot toward a target position, while several methods are investigated for optimizing the intermediate points through which the robot passes [85, 124, 207]. Then, automatic planning methods based on Linear Temporal Logic requirements are presented for a single robot [111] and for a team of robots that has to fulfill a global mission [54, 118]. For e resulting robot trajectories, the chapter includes a collision avoidance strategy using initial time delays [210].

Chapter 6 presents planning methods for a team of identical robots based on Petri net models. The targeted requirement can be expressed through Boolean-based specifications [147, 148, 206] or through Linear Temporal Logic formulas [119, 120], while the methods include integer optimizations. The chapter also presents two specific problems, a sequencing one, where a linear programming optimization suffices to obtain a solution [116], and a task accomplishment one, which needs to gather more robots in the same place [117, 164]. Collision and deadlock avoidance by means of resource allocation techniques are also discussed [123].

Finally, Chapter 7 presents the conclusions of this monograph, as well as a brief discussion about some possible future works in the field of path planning in mobile robotics.

2

Robot Motion Toolbox

2.1 Introduction

Traditionally many courses dealing with mobile robotics are taught through lectures and lab sessions [21, 128, 154, 162]. Lectures focus on introducing and explaining theoretical concepts, e.g., abstract mathematical and physical developments, geometry, algorithms, etc. Lab sessions are related to assembling, programming, and testing mobile robots often using robotic kits like Lego Mindstorms [46, 49]. However, sometimes the transition from theoretical lectures to lab sessions, where actual robots are employed, is not straightforward, especially for introductory courses. In this regard, it is becoming popular to use interactive software simulators to drive the theoretical explanations. It is clear that students will comprehend more quickly a theoretical concept if they can manipulate several parameters and instantly see the effect/result [56, 224]. Thus, if such software offers the possibility to compare similar approaches, the learning process will be even worthier.

In order to overcome the gap between theoretical concepts and lab sessions related to mobile robotics, numerous software simulators have been proposed in the literature and applied in schools, colleges, and universities worldwide; see Figure 2.1 for a quick overview [16, 20, 95]. Many of them have been implemented in Matlab. A pioneering work in the field of robotics simulators was [40], in which Prof. Peter Corke introduced the first popular Robotics Toolbox for Matlab. This toolbox allows the user to work with serial-link manipulators considering kinematics and dynamics. An extended version of this toolbox adding mobile robotics functionalities appears in [41]. Here a comprehensive set of Matlab and Simulink scripts deals with mobile robot navigation: motion planning, motion control, as well as localization and mapping. Another pioneering toolbox specifically related to mobile robotics is SIMROBOT [98], allowing the user to simulate the behavior of one or more mobile robots. Each robot could be equipped with several virtual sensors. The control comes from using Fuzzy Logic and Neural Network toolboxes

Path Planning of Cooperative Mobile Robots Using Discrete Event Models, First Edition.
Cristian Mahulea, Marius Kloetzer, and Ramón González.

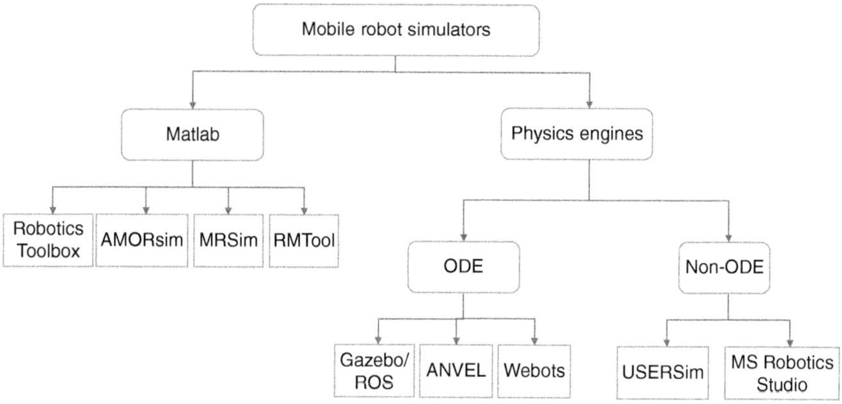

Figure 2.1 Simulators for path planning and motion control. Different simulators for path planning and motion control with their main characteristics.

available in Matlab. However, it constitutes an obsolete tool and no support is available today (it was implemented only for Matlab 5). A more recent simulator extended from SIMROBOT is MRSim [44]. This software is mainly focused on multi-robot simulation. In this sense, interesting features such as inter-robot communication and collective mapping are available. In [166], a Matlab-based simulator called AMORsim is presented. In particular, it deals with localization and permits evaluation of the effects of systematic and non-systematic errors in robot pose estimation. Additionally, obstacle detection is simulated. A navigation toolbox related to the problem of Simultaneous Localization and Mapping (SLAM) is detailed in [8]. In this case, previously recorded sensorial data can be loaded to reproduce a desired simulation.

Other simulators specifically designed for education, but implemented in other programming languages rather than Matlab, are also popular in the literature. For example, the Mobile Robotics Interactive Tool (MRIT) is implemented in Sysquake [89]. MRIT mainly involves motion planning: global path planning algorithms and reactive navigation approaches can be selected for different robot kinematics. In [30], a virtual and remote laboratory based on Easy Java Simulations (EJS) is presented. The main application of this software is to improve the study of sensors in order to obtain a map.

A trending topic in the field of general-purpose robotic simulators comprises the integration of physics engines. Physics engines account for a huge popularity in the video game industry. The main feature of a physics engine and the reason why it is becoming so popular in robotics today is that it can simulate the dynamics involving rigid bodies such as collisions and frictions. One of the most well-known physics engines is the Open Dynamics Engine (ODE) (http://www.ode.org). ODE constitutes an open source, high-performance library for simulating rigid-body dynamics. ODE is integrated within the 3D environment Gazebo, which belongs to the popular Robot Operating System (ROS) framework [173]. Another application of the physics engine ODE is in the commercial robotic simulator ANVEL [172]. Among ANVEL's top features is the capacity to simulate the interaction between the wheels of a robot (friction) with the terrain. This allows the researchers to investigate situations such as sinkage, embedding, and vehicle entrapment [83]. Webots is a ground robot simulator that also uses the ODE engine for its physics simulations [153]. Webots provides a rapid prototyping environment for modeling, programming, and simulating any kind of mobile robot.

An interesting robotic simulator is the open-source urban search and rescue robot simulator (USARSim). This simulator is based on the physics engine Unreal2 [26]. USARsim constitutes the simulation software used to run the Virtual Robots Competition within the Robocup initiative (http://rescuesim.robocup.org). Among the main features, it can simulate multiple sensors and actuators, several platforms (wheeled, legged, and tracked robots), computer vision, and motion controllers. Microsoft's Robotics Studio includes a simulation runtime that is similar to a game engine in terms of physics and visualization capabilities. This software uses the Ageia PhysX physics engine, which is one of the highest fidelity engines available to date (owned by NVIDIA).

The above-mentioned simulators show some limited options for mobile robotics education, especially in introductory courses: (i) very few packages permit extension of their capabilities by adding new functions (scalability); (ii) some of them are standalone applications, so the interested users cannot take advantage of a particular function for their own application (closed implementation); (iii) in order to work with some of them, a reasonable knowledge of the programming environment is required, e.g. Matlab, Simulink, C++; and (iv) some require a complicated setup before carrying out interesting simulations, e.g. libraries, dependences, etc. So the feeling of interactivity can be easily lost, while the student becomes bored, especially if one is not familiar with the programming language or the software suite.

The Matlab-based interactive software tool presented here, named RMTool, is freely available at https://cmahulea.github.io/RobotMotionToolbox-RMTool-under-MATLAB/. The ultimate goal of this software is that students can focus more on concepts such as robot path planning and motion control than on the program implementation (they do not need previous knowledge of Matlab or programming). The proposed toolbox offers an easy to run simulation whatever the tested algorithm. This leads to a high realism in interactivity and students do not have to wait to see the effects of their actions. Additionally, the proposed software simulator comprises the following advantages: (i) scalability: new functions/algorithms can be easily added to this toolbox by just modifying a single file (GUI main file); (ii) visual appearance: we have focused on developing a friendly GUI with just one single window with straightforward and intuitive buttons, icons, and axes objects; (iii) it is open-source so any interested educator or researcher is welcome to use it, and specific functions (scripts) can be employed outside this toolbox; (iv) it can be configured in such a way that a team of robots can work together on a global mission (expressive tasks); and (v) it is implemented in Matlab, which constitutes a widespread and well-known suite in the academic community.

The remainder of this chapter is organized as follows. Section 2.2 details the GUI dealing with the proposed simulator. The path planning algorithms implemented within the robot simulator are briefly reviewed in Section 2.3. Sections 2.4 and 2.5 include the considered robot kinematic models and the motion control algorithms. Some illustrative examples demonstrating the benefits of the proposed simulator are discussed in Section 2.6. Finally, conclusions close the chapter in Section 2.7.

2.2 General description of the simulator

The Robot Motion Toolbox (RMTool) presents a friendly Graphical User Interface (GUI) that enables the user to easily insert the input parameters in order to perform the simulations (Figure 2.2). It has several graphical environments (MATLAB axes objects) where the results of simulations are displayed. The environment where the robots are moving can be either defined by the user in a Matlab axes object or it can be imported from an .mat or an .env file. There are several editable text box objects and menus where the user can specify various parameters for path planning and motion control. As shown in Figure 2.2, the GUI can be divided into the following areas with different functionalities: *Menu Bar* (1), *Drawing Area (Trajectory/Workspace)* (2), *Path Planning Panel* (3), *Mission Type Panel* (4), and *Simulation Panel* (5).

(1) Menu Bar Panel (placed horizontally, on top of the main window of the RMTool) displays a set of five drop-down menus: *File*, *Setup*, *Environment*, *Path Planning*, and *Help*.

- The *File* menu offers facilities for file-handling operations and contains the following three options:
  - *Open*: allows a user to load previously used data.
  - *Save*: saves the current data in a .mat file (which can be later loaded with the previous option).
  - *Export as eps*: saves the current figure as an .eps file.
- The *Setup* menu deals with different tuning parameters and has the following options (see Figure 2.3):

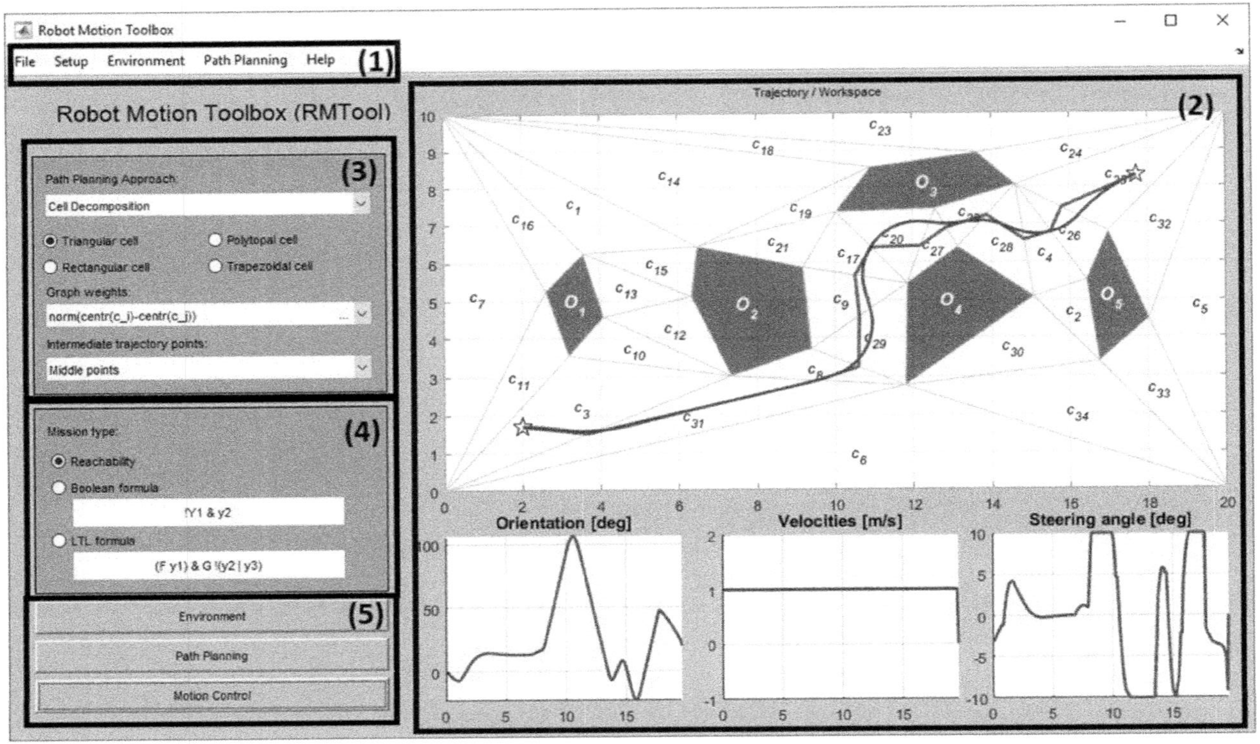

Figure 2.2 Main window of Robot Motion Toolbox (RMTool).

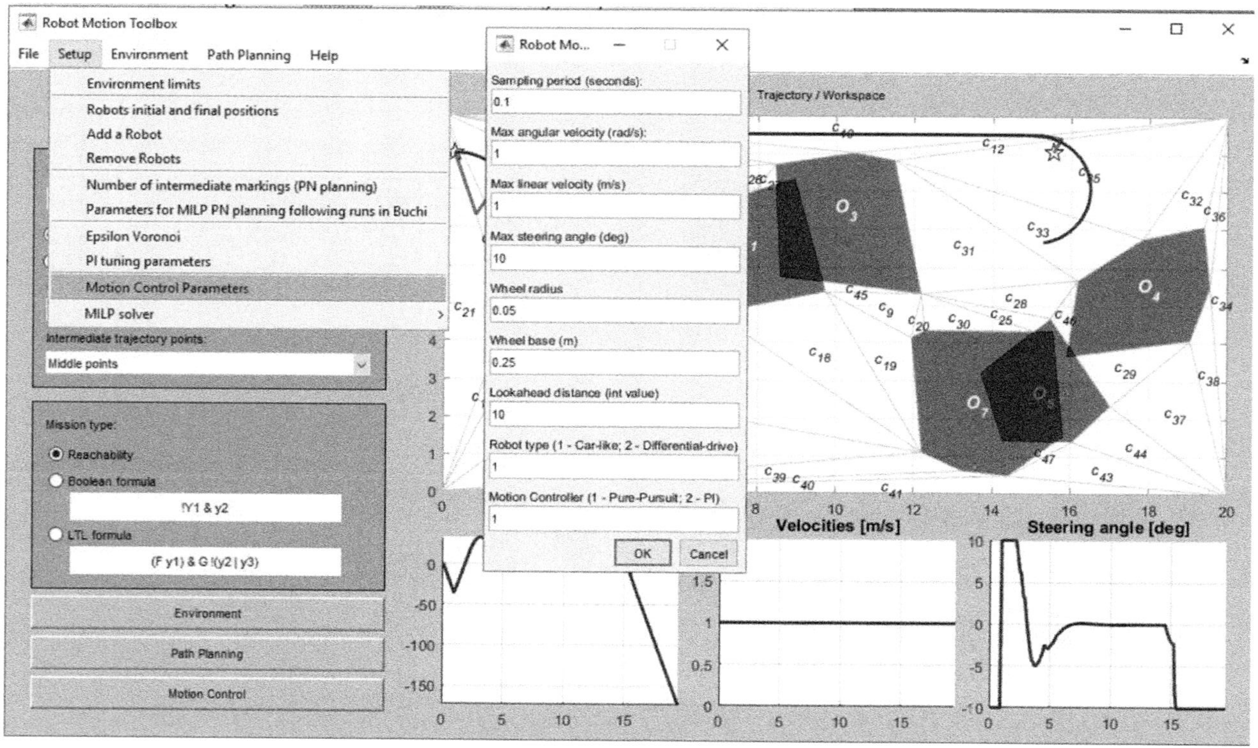

Figure 2.3 Submenu with information about the configuration of the robot and the motion controllers.

- *Environment limits* allows the user to change the limits of the axes object placed on the top-right part of the GUI containing the representation of the environment.
- *Robots initial and final positions* can be used for changing the start and the goal positions of the robots.
- *Add a robot* and *Remove robots* allow the user to add or remove a robot from the environment.
- *Number of intermediate markings* is a variable dealing with the number of intermediate markings for the path planning with the Petri net approach, in order to avoid spurious markings (see Chapter 4).
- *Parameters for MILP PN planning following runs in Büchi*: permits the user to specify different parameters related to the optimization problems in the case of path planning with Petri net models following runs in Büchi automaton (see Section 6.3).
- *Epsilon Voronoi* is a tuning parameter of the Voronoi algorithm.
- *PI tuning parameters* for setting the parameters of the PI controller.
- *Motion Control Parameters* for setting the main variables for the motion controllers and the kinematics of the robot.
- *MILP solver* with the three libraries used for solving the optimization problems, i.e. CPLEX [102], GLPK [150], and Intlinprog [196] (Matlab built-in function, for versions after 2014).

• The *Environment* menu is related to loading or saving data related to the environment and has the following submenus:
  - *Load environment* allows the user to load a configuration of the environment (set of obstacles and robot starting and goal positions) from a previous simulation.
  - *Save environment* allows the user to save the environment for a later simulation.
  - *Export to workspace* exports the current data regarding the environment and/or the Petri net model to the Matlab workspace.
  - *Export to figure window*: the user can save the current environment into a new single figure.
• *Path Planning* menu permits the user to save the results after applying a path planning algorithm.
  - *Export to workspace* saves the simulation results to some variables of the environment.
  - *Export to figure window* saves the results to a new Matlab figure object.
• *Help* menu provides credential information.

(2) Drawing Area Panel (Trajectory/Workspace). This panel is composed of four axes objects used to represent the evolution of different variables and the environment. The main axes object placed on the top right side of the main window is used to define the obstacles, to represent the trajectory obtained by path planning algorithms and to display the simulation of the controlled motion. On the bottom part of the main axes there are three smaller axes objects used to represent the control actions after a simulation: orientation, linear velocity and steering angle of the robot.

(3) Path Planning Panel. It is placed in the left-top side of the GUI and consists of three drop-down menus and four radio button objects. By using the first drop-down menu, the user can select the approach used for path planning. The RMTool implements three approaches: cell decomposition, visibility graph and generalized Voronoi diagram. If cell decomposition is selected, the user can choose the type of decomposition by using the radio buttons. Moreover, since in this case the trajectory will rely on a graph search, the second drop-down menu permits the selection of weights in the graph corresponding to cell decompositions. The third drop-down menu allows the user to specify the method for generating the passing points between adjacent cells from the sequence of regions that should be followed by the robot. The following options are available: middle points (no optimization) or norm 1 and 2, infinity or MPC (by solving an optimization problem, as detailed in Section 5.2).

(4) Mission Type Panel. This part of the tool can be clearly divided into two functionalities. The first part called "Reachability" runs this toolbox for a single-robot and a standard navigation problem (reaching a desired point/region by avoiding the obstacles). That is, the user can play with all the previous options for one robot only.

The second part is devoted to path planning using expressive or high-level tasks, mainly for a group of identical robots. Imagine the situation where a team of robots should perform some motion tasks such as visiting a region of interest before visiting another one or avoiding a particular region. The theoretical concepts behind this part of the tool is related to the case when the mission of the team is expressed as Boolean or Linear Temporal Logic (LTL) formulas.

(5) Simulation Panel consists of three push buttons that are used to run different options. The *Environment* button starts the routine for defining new obstacles or regions of interest in the environment. A new Matlab window is opened and the number of new obstacles and eventually the number of robots (if high-level tasks are considered) should be introduced. By using the mouse the user can introduce the obstacles or regions of interest in the main axes object. Those objects can be placed either manually or randomly. The *Path Panning* button starts automatically the path planning approach (selected in the *Path Planning Panel*) and draws (in the main axes object of the *Drawing Area Panel*) the trajectory that the robot should follow. Finally, the *Motion Control* button simulates the trajectory followed by using the robot model and control parameters defined in the submenu *Motion*

Control Parameters in the menu *Setup*. The resulting robot trajectory is displayed in the main axes object of the *Drawing Area Panel*, while graphical representations of control actions are represented in the other axes objects.

2.3 Path planning algorithms

The path planning algorithms implemented in RMTool considers fully actuated point mobile robots. As explained in Chapter 1, the path planning problem refers to finding a path (reference trajectory) from a given initial position to a desired goal position or finding a path that accomplishes a given high-level mission. This reference path must avoid any obstacle in the environment. The path generally deals with minimum-length routes, but in some occasions the shortest distance does not guarantee reaching the goal point, for instance, due to terrain unevenness [101], or constrained routes where the robot must pass through some waypoints [149].

Commonly, in order to accomplish the path planning objective, a (geometric) map of the environment is required [58, 191]. Depending on the considered geometrical map, two general strategies can be identified [191]:

- Cell decomposition - partition the free environment space. In particular, rectangular, triangular, polytopal, and trapezoidal decompositions are employed in the actual version of RMTool.
- Road map - identify a set of fast routes within the free space. Visibility graphs and generalized Voronoi diagrams constitute two examples.

The geometric map represents a finite abstraction of the environment (either via a set of cells or a set of fast routes). Once the geometric map is available, planning strategies require some form of graph searching algorithm in order to identify the path from the initial position to the goal. The toolbox presented in this chapter uses Dijkstra's algorithm, which is generally employed in shortest-path graph searches [58]. In situations when a particular heuristic is of interest, optimal route algorithms such as A^* or E^* may be preferred [58, 168], but this is not the case for the currently assumed setup.

Cell decomposition techniques partition the free space (not covered by obstacles) into a set of regions having the same geometrical shape. RMTool includes decompositions of environments with polygonal and convex obstacles into cells having triangular, trapezoidal, polytopal or rectangular shape. More details on the software implementation of decomposition routines and supporting algorithms will be described later in Chapter 3.

A finite graph is constructed by assigning each cell to a node, transitions based on adjacency between cells, and graph weight computation chosen by the user via a drop-down menu. The Dijkstra algorithm is run on this finite abstraction of the free space, with start and goal nodes corresponding to cells that include the initial and final robot positions. The output is a sequence of cells to be followed, and any path belonging to this sequence is valid since it avoids obstacles.

A piece-wise linear robot trajectory is constructed by linking the start point with a point on the line segment shared by the first and second cells, and so on until the last cell. The points on the line segments shared by two successive cells can be chosen to be the middle point of the segment or can be computed via an optimization problem. The convex optimization problem minimizes a sum of norms of all linear segments of the obtained path, and our toolbox supports norm one, two, or infinity (see Chapter 3 for a further description).

Visibility graph it is a well-known approach solving a minimum-length path by following a sequence of edges (straight lines). These line segments connect a set of obstacle vertices visible one from the other, as well as the start and goal points [35, 191]. The V-graph has nodes corresponding to vertices and edges corresponding to line segments, while each arc cost is given by the traveled distance along each line segment. The Dijkstra's graph search algorithm yields the shortest path from the start to the goal, but the reference trajectory in close to obstacles, since it contains some obstacle vertices.

Recall that the robot is represented by a point and is capable of omnidirectional motion. One mechanism for compensating this assumption is to consider that the robot has a circular cross-section and to inflate (dilate) all obstacles in the environment by an amount equal to the robot's radius [58, 132].

Generalized Voronoi diagram. Contrary to the V-graph planner, the algorithm based on the Generalized Voronoi Diagram (GVD) tends to maximize the distance between the robot and obstacles in the map. In particular, for each point in the free space, the distance to the nearest obstacle is computed. The GVD contains only equidistant points from two or more closest obstacles, and this leads to straight and parabolic segments (when polygonal obstacles are considered) [35, 135]. As in the previous case, once the graph with the set of paths maximizing the clearance between points and obstacles is calculated, the graph search algorithm returns to the reference route. The resulting robot's path accesses the GVD from the initial position, follows some of its segments, and leaves the GVD near the goal point.

2.4 Robot kinematic models

In the context of robotics, a model is defined as a set of mathematical differential equations that represents the behavior of a robot. In this sense, kinematic and dynamic modeling constitutes a key issue that can be employed for estimating robot location and implementing software simulators, among other issues.

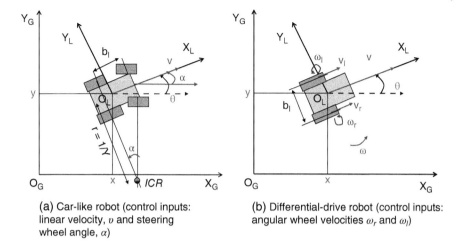

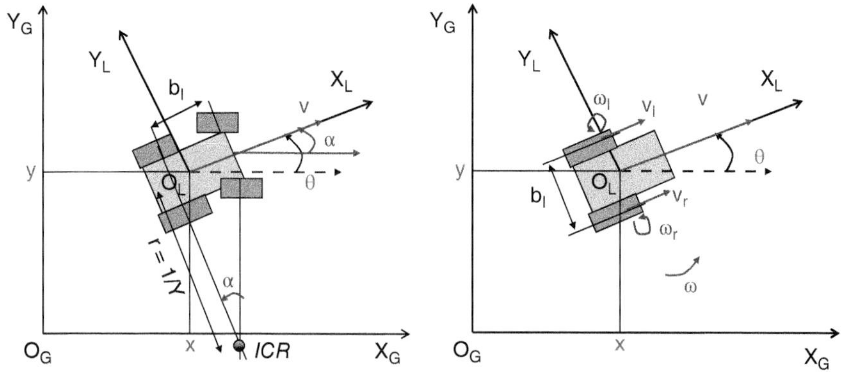

(a) Car-like robot (control inputs: linear velocity, v and steering wheel angle, α)

(b) Differential-drive robot (control inputs: angular wheel velocities ω_r and ω_l)

Figure 2.4 Robot configurations considered in the RMTool software.

RMTool implements two kinematic models in order to simulate the motion of a car-like robot and a differential-drive robot. Furthermore, such models will be employed for estimating the robot position at each sampling instant (integrated from the initial robot position). This position will be feed back to the selected motion controller in order to ensure path following. Notice that car-like and differential-drive models have been selected because they are the most common configurations in wheeled/tracked mobile robots [58, 81, 191].

Car-like robot. Four-wheeled car-like robots (Ackerman vehicles) are popular in the field of mobile robotics [191], and this configuration was one choice to include in RMTool. In general, such a robot is modeled in terms of a bicycle model. The bicycle has a rear wheel fixed to the body and the plane of the front wheel rotates about the vertical axis to steer the vehicle (Figure 2.4a) [41]. The vehicle is assumed to only move forward, and the wheels cannot slip sideways.

The car-like configuration leads to circular paths when the vehicle turns; these circular arcs are centered at a point known as the Instantaneous Centre of Rotation (ICR). Therefore, the angular velocity is given by

$$\dot{\theta}(t) = \omega(t) = \frac{v(t)}{r(t)}, \tag{2.1}$$

where θ is the vehicle orientation with respect to the x-axis, v is the forward speed of the robot (imposed by the real wheels), while r is the turning radius. This turning radius is defined as

$$r(t) = \frac{b_l}{\tan \alpha(t)}, \tag{2.2}$$

where $b_l \in \mathbb{R}^+$ is the length of the vehicle or wheel base,[1] while $\alpha \in \mathbb{R}$ is the steering angle that is mechanically constrained and its maximum value influences robot motion. Notice that this important feature can be analyzed with the software tool proposed here because the user can change the maximum angle that the steering wheel can turn (see Section 2.6 for illustrative examples about this aspect).

Based on Figure 2.4a, one can write the following equations dealing with the car-like robot kinematics [135]:

$$\dot{x}(t) = v(t) \cdot \cos\theta(t),$$
$$\dot{y}(t) = v(t) \cdot \sin\theta(t),$$
$$\dot{\theta}(t) = \frac{v(t)}{b_l} \cdot \tan\alpha, \tag{2.3}$$

where $[x \quad y \quad \theta]^T \in \mathbb{R}^2 \times [0, 2\pi)$ represents the pose (position and orientation) of the mobile robot and $t \in \mathbb{R}$ is the time variable. Notice that in Figure 2.4, the symbols X_G, Y_G, and O_G represent the inertial (or global) Cartesian coordinate system, while X_L, Y_L, and O_L represent the local coordinate system of the robot. More information can be found in [81].

Differential-drive robot. Another used configuration in mobile robotics is the differential-drive one. In fact, this is the easiest mechanical configuration because only two parallel driving wheels mounted on an axis are enough to move the robot (Figure 2.4b). In particular, forward spin of the right wheel (v_r) and a stopped left wheel ($v_l = 0$) result in counterclockwise rotation around the point where the left wheel touches the ground, while equal wheel speeds result in straight motion. Different combinations of v_r and v_l yield instantaneous rotations around points on the line segment linking the contact points between wheels and the ground surface.

As known, in the absence of wheel slip, the linear velocity of the wheels is given by [191]

$$v_r(t) = \rho \cdot \omega_r(t),$$
$$v_l(t) = \rho \cdot \omega_l(t), \tag{2.4}$$

where v_r and v_l are the linear velocities, while $\omega_r, \omega_l \in \mathbb{R}$ are the angular velocities of the right and left wheels, respectively, and ρ is the wheel radius. The linear velocity of the robot is given by

$$v(t) = \frac{v_r(t) + v_l(t)}{2}. \tag{2.5}$$

1 In this book we assume square robots for which the vehicle length is equal to the wheel base.

In this way, the kinematic model for a differential-drive mobile robot is given by [22, 191]

$$\begin{aligned}\dot{x}(t) &= \frac{v_r(t) + v_l(t)}{2} \cdot \cos\theta(t),\\ \dot{y}(t) &= \frac{v_r(t) + v_l(t)}{2} \cdot \sin\theta(t),\\ \dot{\theta}(t) &= \frac{v_l(t) - v_r(t)}{b_l},\end{aligned} \tag{2.6}$$

where $b_l \in \mathbb{R}^+$ is the distance between the wheel centers equal to the length of the robot by assumption.

2.5 Motion control algorithms

Mobile robots must have effective motion controllers that generate proper control actions to successfully steer the robot such that it follows the reference trajectory given by the path planner. Closed-loop motion controllers use the current robot position to make new decisions (control inputs) that will eventually drive the robot to a desired goal or waypoint [80, 81, 191, 202]. Recall from Chapter 1 that there are three main control problems in the field of mobile robotics: moving to a point, path following, and trajectory tracking. RMTool considers the path following control problem. In this case, a reference point on the robot must follow a path in the Cartesian space $(x^{ref}(t), y^{ref}(t))$ starting from a given initial configuration. This path may come from a sequence of coordinates generated by the path planner, as in Section 2.3. Examples of trajectory following approaches are the Pure Pursuit algorithm [4], reviewed in Subsection 2.5.1, and a form of Proportional Integral (PI) controller introduced in [41] and explained in Subsection 2.5.2.

As previously mentioned, a robot with a specific kinematic model as in Section 2.4 may sometimes not be able to closely follow a reference path designed for a generic robot as in Section 2.3. If the simulated trajectory leaves the environment or hits an obstacle, the user is informed.

2.5.1 Pure pursuit algorithm

The Pure Pursuit is a well-known strategy for the mobile robotics community. It was formulated in the framework of the CMU Terregator and Navlab projects [4, 45]. The idea behind the Pure Pursuit control law is based on repeatedly fitting different circular arcs (geometrical approach) to different waypoints as the robot moves forward until the final goal point is reached. These waypoints are obtained as a constant lookahead distance (along trajectory) away from the closest point between the current robot position and the reference path (Figure 2.5). The output

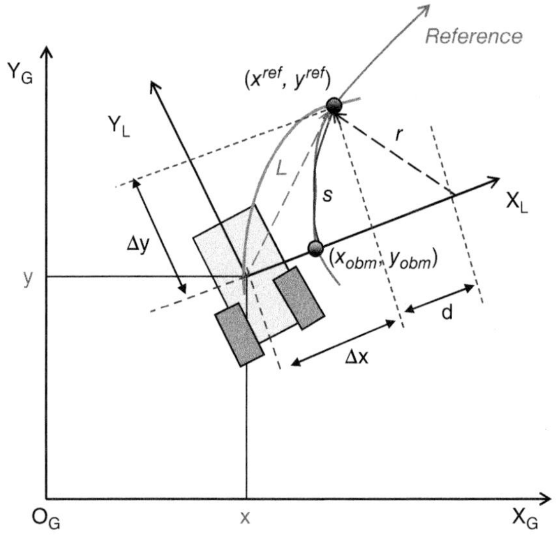

Figure 2.5 Pure Pursuit strategy.

of the controller is the desired steering angle of the driving wheel of the robot (in the case of a car-like robot) or the wheel velocities (for differential-drive robots), while the robot's linear velocity is constant.

From Figure 2.5, the following mathematical equations hold [45]:

$$r(t) = \Delta x(t) + d(t), \tag{2.7}$$

$$r(t)^2 = d(t)^2 + (\Delta y(t))^2, \tag{2.8}$$

solving for the turning radius, r,

$$r(t)^2 = (r(t) - \Delta x(t))^2 + (\Delta y(t))^2, \tag{2.9}$$

leads to

$$r(t) = \frac{(\Delta x(t))^2 + (\Delta y(t))^2}{2\Delta x(t)}. \tag{2.10}$$

Finally, the curvature that the robot must follow is

$$\gamma(t) = \frac{1}{r(t)} = \frac{2\Delta x(t)}{L^2}. \tag{2.11}$$

where r is the radius of the curve from the nearest position of the robot and the reference path (x_{obm}, y_{obm}) to the next reference point (x^{ref}, y^{ref}). The variables Δx, Δy represent the longitudinal and lateral error between the current position of the robot and the desired position. Notice that the Pure Pursuit is a proportional controller of the steering angle using the x-displacement as the error and $2/L^2$ as the gain. In this sense, tuning the lookahead distance, L, plays a major role in the performance of the controller. In Section 2.6 some examples highlight the significance of the lookahead distance.

Once the desired curvature $\gamma(t)$ is computed by Eq. (2.11), the control actions to be used in Eq. (2.3) for a car-like robot are $v(t) = constant$ and $\alpha = atan\left(b_l \cdot \gamma(t)\right)$ (from Eq. (2.2)). For a differential-drive robot the control action to be used in (2.6) are computed as follows. Using the last equation in (2.6) and Eq. (2.1), we can write

$$\frac{v_l(t) - v_r(t)}{b_l} = \frac{v(t)}{r(t)} = \gamma(t) \cdot v(t),$$

and considering also Eq. (2.5), the following actions are obtained:

$$v_r(t) = v(t) - \frac{b_l \cdot v(t) \cdot \gamma(t)}{2}$$

and

$$v_l(t) = v(t) + \frac{b_l \cdot v(t) \cdot \gamma(t)}{2}$$

where $v(t)$ is a constant speed.

The function that has been implemented within RMTool is called: *rmt_pure_pursuit*. This is the syntax of this function: *[array_time, array_gamma, array_v, array_pos] = rmt_pure_pursuit(input_variables, ref_trajectory, orientation, obstacles, Nobstacles)*, where

- *array_time*: vector with the time required by the robot to reach the goal position.
- *array_gamma*: vector with the sequence of steering angles.
- *array_v*: vector with the linear velocity of the robot (which is a constant value).
- *array_pos*: vector with the positions and orientations followed by the robot (trajectory followed).
- *input_variables*: variables dealing with the sampling period, the position and size of the robot, the velocities, the type of robot, and the limits of the environment.
- *ref_trajectory*: reference trajectory (obtained after applying a path planning algorithm).
- *orientation*: initial orientation of the robot.
- *obstacles*: variables defining the position and size of the obstacles.
- *Nobstacles*: number of obstacles defined by the user.

2.5.2 PI controller

For comparison purposes, RMTool also includes the path following algorithm implemented in [41], where a Proportional Integral (PI) controller is employed for controlling the robot position. Again, the robot maintains a distance behind a given waypoint. First, the Euclidean distance between the current robot position and the desired waypoint is calculated as

$$e_d(t) = \sqrt{(x^{ref} - x(t))^2 + (y^{ref} - y(t))^2} - d^*. \tag{2.12}$$

where x^{ref} and y^{ref} are related to the reference trajectory, $x(t)$ and $y(t)$ are related to the position of the robot, and d^* is a constant point ahead of the robot on the trajectory (similar to the lookahead distance of the Pure Pursuit controller).

The goal is to regulate (2.12) to zero by controlling the robot's velocity using a PI controller

$$v(t) = K_v \cdot e_d(t) + K_i \cdot \int e_d(t) \cdot dt, \tag{2.13}$$

where $v(t)$ is the control input dealing with the linear velocity of the robot. Notice that the integral term avoids an offset error. On the other hand, a second controller steers the robot towards the target, which is at the relative angle

$$e_\theta(t) = \tan^{-1}\left(\frac{y^{ref} - y(t)}{x^{ref} - x(t)}\right). \tag{2.14}$$

Then a simple proportional controller turns the steering wheel of a car-like robot for driving the robot towards the target, that is,

$$\alpha(t) = K_h(e_\theta(t) \ominus \theta(t)), \quad K_h > 0, \tag{2.15}$$

where α is the control input dealing with the steering angle of the robot, K_h is a constant, $e_\theta(t)$ was defined in Eq. (2.14), and $\theta(t)$ is the robot orientation. Furthermore, the symbol $\ominus$ is for difference between angles, keeping the result in $[0, 2\pi)$.

The control actions are computed in a similar way as in the case of the Pure Pursuit controller. For a car-like robot, the control action α from Eq. (2.15) and control action v from Eq. (2.13) are used for a model in Eq. (2.3). For a differential drive robot, both control actions are used to compute the left and right velocities as follows. Using the last equation in (2.6) and Eq. (2.1), we can write

$$\frac{v_l(t) - v_r(t)}{b_l} = \frac{v(t)}{r(t)} = \frac{v(t) \cdot \tan \alpha(t)}{b_l},$$

and considering also Eq. (2.5), the following actions are obtained

$$v_r(t) = v(t) - \frac{v(t) \cdot \tan \alpha(t)}{2}$$

and

$$v_l(t) = v(t) + \frac{v(t) \cdot \tan \alpha(t)}{2}.$$

It is important to point out that the performance of the PI controller depends highly on the values of the gains K_v, K_i, K_h (tuning). In this sense, this software tool permits such values to be adjusted, and hence the user can understand in a better way how they affect the trajectory that follows.

Notice that the main difference between Pure Pursuit and this PI-based controller is that the latter control approach produces the control action for both the traction and steering wheels. In contrast, for the Pure Pursuit approach the linear velocity of the robot is manually fixed and does not change along the experiment. As shown in Section 2.6, this software tool allows an evaluation to be made of the performance of fixing the linear velocity of the robot (Pure Pursuit) against being able to change that velocity depending on the Euclidean error (PI controller).

The function that has been implemented within RMTool is called: *rmt_pi_controller*. This is the syntax of this function: *[array_time, array_gamma, array_v, array_pos] = rmt_pi_controller(input_variables, ref_trajectory, pi_tuning, obstacles, Nobstacles)*, where

- *array_time*: vector with the time required by the robot to reach the goal position.
- *array_gamma*: vector with the sequence of steering angles.
- *array_v*: vector with the linear velocity of the robot.
- *array_pos*: vector with the positions and orientations followed by the robot (trajectory followed).
- *input_variables*: variables dealing with the sampling period, the position and size of the robot, the velocities, the type of robot, and the limits of the environment.
- *ref_trajectory*: reference trajectory (obtained after applying a path planning algorithm).
- *pi_tuning*: variables related to the tuning properties of the PI controller (proportional gain and integral time).
- *obstacles*: variables defining the position and size of the obstacles.
- *Nobstacles*: number of obstacles defined by the user.

2.6 Illustrative examples

2.6.1 Examples about path planning aspects

The examples shown in Figure 2.6 address all the steps related to the application of the path planning algorithms implemented in this software: cell decomposition, visibility graph, and generalized Voronoi diagram. The trajectory obtained

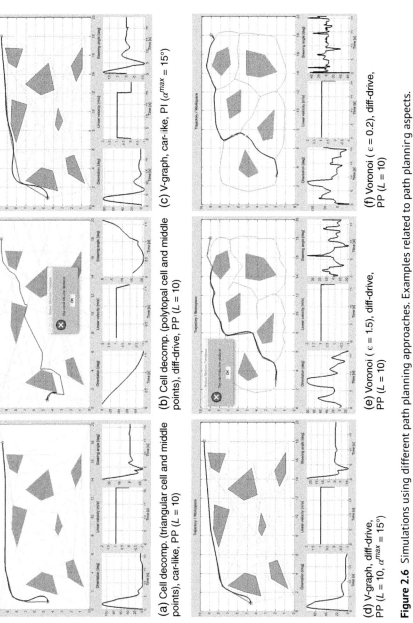

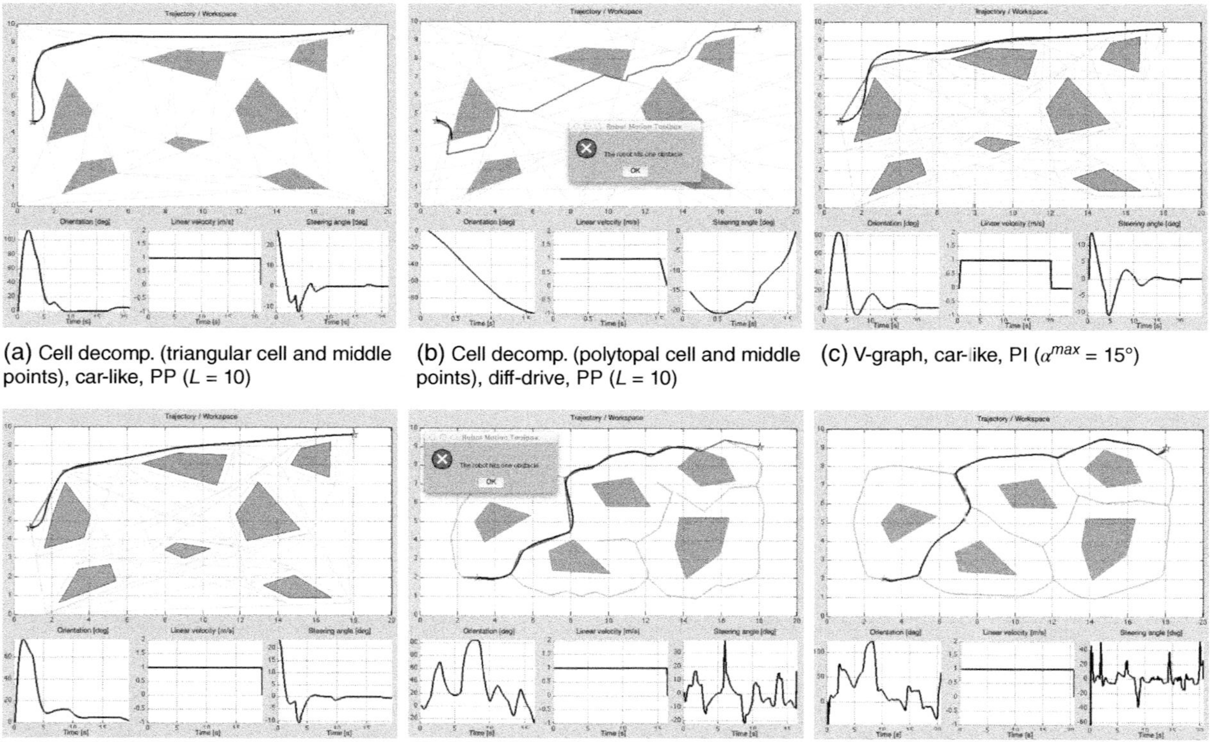

(a) Cell decomp. (triangular cell and middle points), car-like, PP ($L = 10$)

(b) Cell decomp. (polytopal cell and middle points), diff-drive, PP ($L = 10$)

(c) V-graph, car-like, PI ($\alpha^{max} = 15°$)

(d) V-graph, diff-drive, PP ($L = 10$, $\alpha^{max} = 15°$)

(e) Voronoi ($\epsilon = 1.5$), diff-drive, PP ($L = 10$)

(f) Voronoi ($\epsilon = 0.2$), diff-drive, PP ($L = 10$)

Figure 2.6 Simulations using different path planning approaches. Examples related to path planning aspects.

while using a path planning algorithm is shown in red, the actual path followed by the mobile robot is in black, obstacles are in blue, and the grey segments represent the different cells in the cell decomposition approach, the different segments connecting the edges in case of V-graph, and the equidistant routes in the case of Voronoi. The tracking of this trajectory is performed using the Pure Pursuit (PP) method, except in Figure 2.6c where the Proportional Integral (PI) controller was employed. In these experiments the robot always starts with orientation $\theta = 0$.

Figures 2.6a and b deal with cell decomposition using two different approaches to find the reference path: triangular versus polytopal cell decomposition. RMTool allows us to observe a possible outcome of the polytopal approach, where the returned trajectory from path following may be too close to obstacles. Notice that RMTool detects this situation and remarks it to the user via a popup window.

Figures 2.6c and d are related to the visibility graph approach. In the first experiment the car-like model is used, where the steering angle was constrained to 15 [^{o}]. In the second example, where a differential-drive robot is used, the steering angle does not influence the robot motion because the differential-drive configuration is not affected by a steering mechanism (recall that the differential-drive mechanism achieves robot turning by changing the relative velocity of the driving wheels).

Finally, the generalized Voronoi diagram is employed in Figures 2.6e and f. In particular, with RMTool we can observe the importance of a parameter dealing with the interpolation between two consecutive points, that is, the parameter ϵ. The smaller the value of ϵ, the smoother the output will be, but the more computation time it will take. Notice that in the first experiment a high value leads to unconnected segments, due to not enough points having been considered. On the contrary, Figure 2.6f shows a continuous trajectory that is safely followed by the robot, while in Figure 2.6e the outcome of trajectory following was unsatisfactory.

2.6.2 Examples about motion control aspects

The examples shown in this section allow us to easily and visually understand the main properties of motion control and the importance of tuning the controller properly. For that reason, we focus mainly on the lower part of the GUI, that is, the plots dealing with orientation, linear velocity, and steering angle.

Thanks to Figures 2.7a,b and c we may understand the importance of the lookahead parameter in the Pure Pursuit approach. In particular, a higher value ($L = 10$) leads to a smoother trajectory than a smaller value ($L = 2$). This is especially noticeable in the orientation and steering wheel angle plots. Notice the oscillatory behavior after 10 seconds in the steering wheel angle in Figure 2.7b. This fact is also observed in Figure 2.7c. In this case, sudden changes in the robot orientation are noticed from the beginning of the experiment. Such sudden changes are

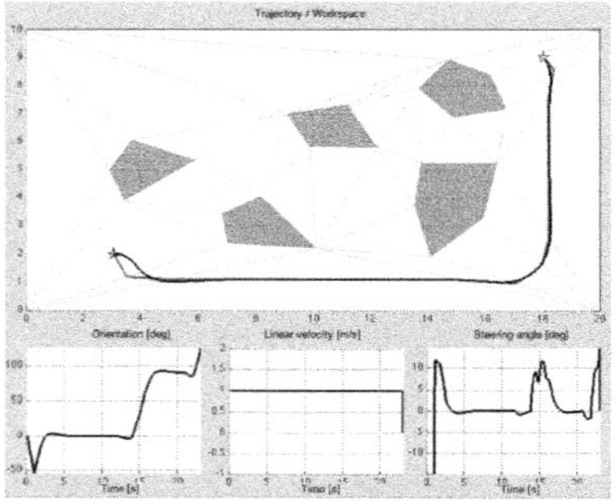

(a) Cell decomp. (tri. cell and mid. points), car-like, PP ($L = 10$, $\alpha^{max} = 15°$)

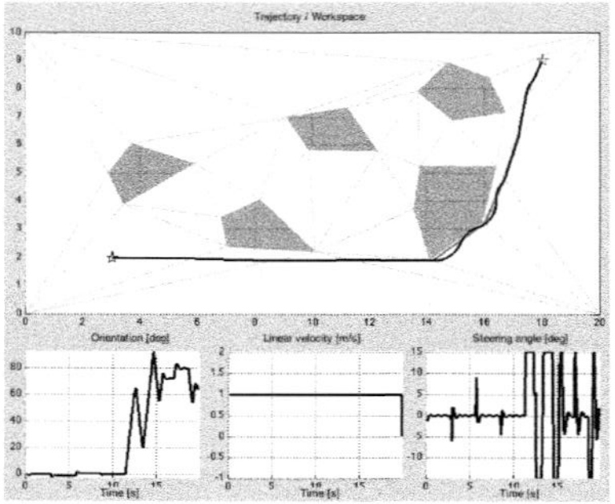

(b) Cell decomp. (tri. cell and norm-2), car-like, PP ($L = 2$, $\alpha^{max} = 15°$)

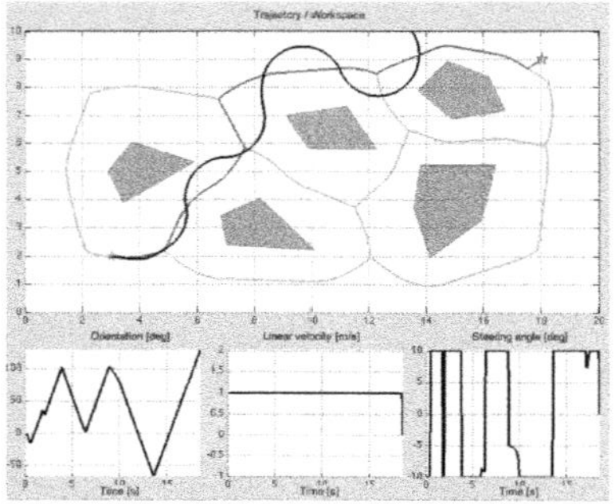

(c) Voronoi, car-like, PP ($L = 2$, $\alpha^{max} = 10°$)

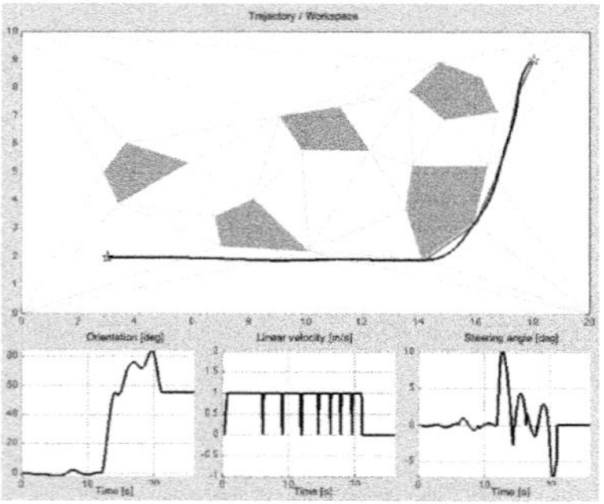

(d) Cell dec. (tri. cell and norm-2), car-like, PI ($\alpha^{max} = 15°$)

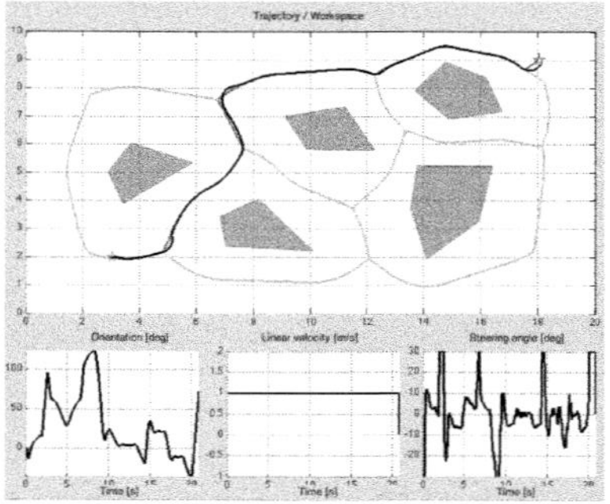

(e) Voronoi, car-like, PP ($L = 10$, $\alpha^{max} = 30°$, $v^{max} = 1$)

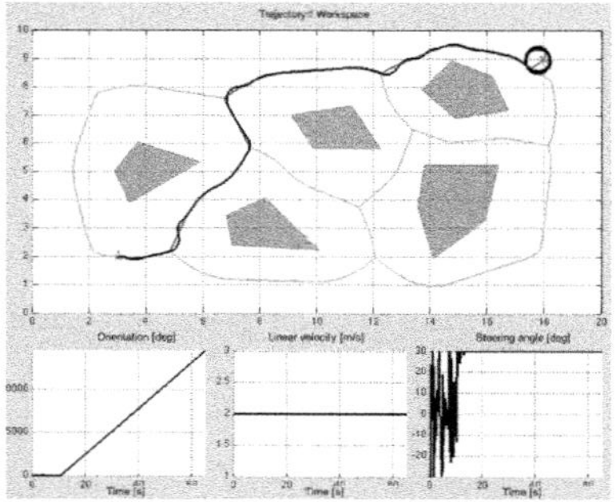

(f) Voronoi, car-like, PP ($L = 10$, $\alpha^{max} = 30°$, $v^{max} = 2$)

Figure 2.7 Simulations using different motion control approaches. Examples related to motion control aspects: performance of the Pure Pursuit (PP) and Proportional Integral (PI) controllers in terms of maximum steering angle (α^{max}), maximum linear velocity (v^{max}), and lookahead distance (L).

produced by the incorrect control actions (see the plot dealing with steering wheel angle).

Figure 2.7d addresses the performance of the PI controller versus PP approach that was used for following the same reference path in Figure2.7b. RMTool permits us to understand the main difference between these controllers that occurs in the case of PP where the linear velocity is always fixed no matter the circumstances. In the case of the PI approach linear velocity depends on the longitudinal error and hence it varies along time. The importance of this fixed linear velocity in the Pure Pursuit approach is clearly shown in Figures 2.7e and f. Observe that in the first plot, with a fixed linear velocity of 1 [m/s], the robot closely follows the reference. On the contrary, when the linear velocity is increased to 2 [m/s], the robot cannot reach the goal point.

2.6.3 Examples about multi-robot systems and high-level tasks

This last set of examples makes use of the RMTool's capabilities for defining high-level tasks such as visiting certain regions of interest or avoiding others through the use of logical (and temporal) operators. We consider in both cases the same environment with two robots and six regions of interests denoted as O_1, O_2, O_3, O_4, O_5, and O_6.

Figure 2.8 represents the case where the formalism LTL is used to define the high-level task to be accomplished by a group of robots. In this case, the desired mission is stated according to the following formula:

$$\phi = F(y2)\&F(y3)\&G(!y4)\&(!y2Uy1)\&F(y5\&y6)$$

which means:

- Regions O_2 and O_3 have to be eventually visited.
- Region O_4 should be avoided at all times.
- Region O_2 can be visited only after O_1 was visited.
- The (disjoint) regions O_5 and O_6 should be eventually occupied at the same time (by both robots).

Figure 2.9 represents the case where the Boolean formula is used to define the task to be accomplished by the group of robots. The goal of this experiment is to see that the path planner changes the mission of the robots according to the task and to the distance between the robots and those regions of interest to be visited or avoided. That means that the global distance traversed by the robots is the shortest one possible. Observe that in this case the robot 2 reaches three regions (O_3, O_4, O_5) and robot 1 only visits one region (O_1).

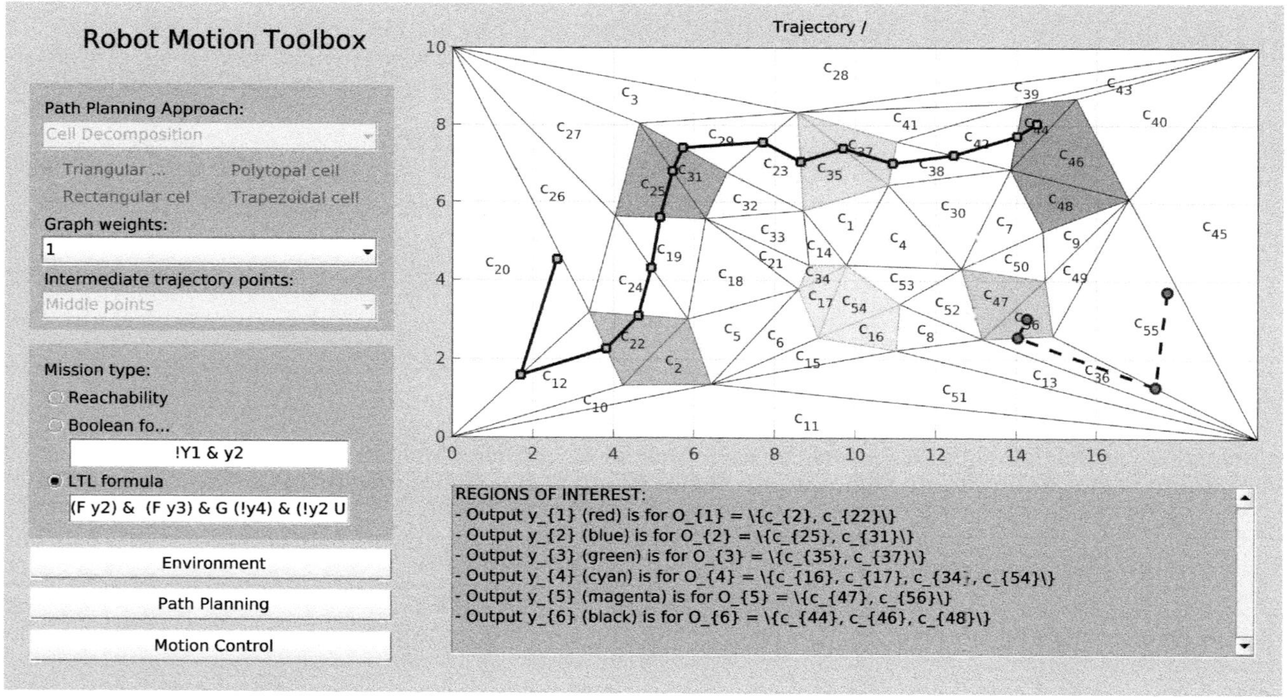

Figure 2.8 Examples related to multi-robot systems and the LTL task by using RMTool and Petri net models. Two robots initially located in c_{20} and c_{55} that should fulfill the following LTL task $F(p2)\& \ F(p3)\& \ G(!p4)\& \ (!p2Up1)\& \ F(p5\&p6)$.

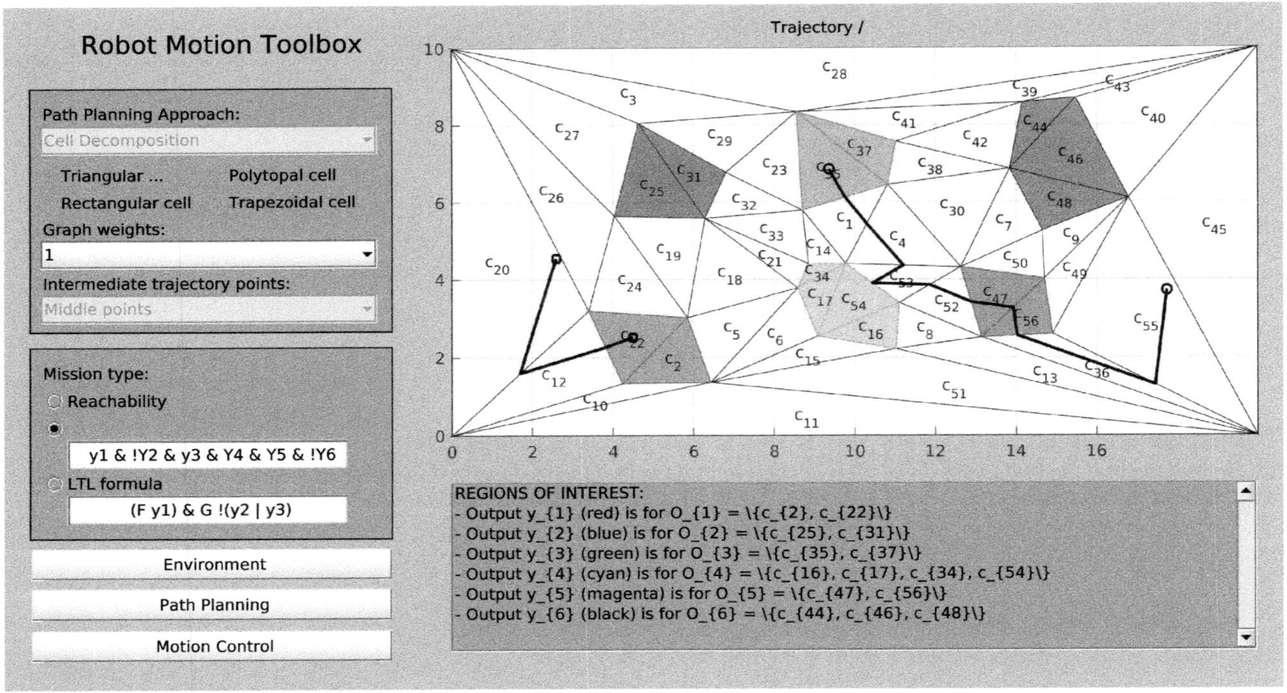

Figure 2.9 Examples related to multi-robot systems and the Boolean task by using RMTool and Petri net models. Two robots initially located in c_{20} and c_{55} that should fulfill the following Boolean task $y_1 \& !Y_2 \& y_3 \& Y_4 \& Y_5 \& !Y_6$, meaning that O_1 and O_3 should be reached in the final state, O_4 and O_5 should be crossed during the robot motion, while O_2 and O_6 should never be crossed.

2.7 Conclusions

According to our professional experience teaching mobile robotics in several universities, the use of software simulators during theory lessons constitutes a key element. Even before application hours, it constitutes an excellent idea in order to quickly review theoretical concepts. This chapter introduces RMTool, an interactive tool aimed at helping researchers and students to understand basic concepts dealing with ground mobile robots. Among other issues, RMTool allows the student to clearly understand the importance of path planning and motion control. As shown through some examples, the choice of a path planner can lead to trajectories closer to obstacles, shorter paths, or longer paths but with larger distance from obstacles.

With the proposed software simulator, a professor can easily illustrate the influence of a bad tuned motion controller. As shown with the pure pursuit approach, the values of the lookahead distance and the fixed linear velocity can either lead to closely following the reference path or can cause undesirable outcomes. Finally, the user can run different mission tasks (visiting certain regions and/or avoiding others) for multi-robot systems via symbolic formalisms such as LTL and Petri nets.

3

Cell Decomposition Algorithms

3.1 Introduction

Different partitioning techniques found applicability in constructing solutions to problems from various domains, such as discrete and hybrid systems [91, 121, 198], image segmentation [185], computational geometry [15], and robotics [35, 135]. In this chapter, we use such techniques in problems dealing with mobile robots, where the common name for an involved partition is "cell decomposition" [35, 132, 135].

Cell decompositions allow the abstraction of an initial continuous-space problem into a finite state representation. The main idea is to decompose a given bounded environment into a set of non-overlapping regions having the same shape. The union of these regions (called cells) should cover or approximate a subset of interest from the environment, e.g. the space not occupied by obstacles. Once a decomposition is obtained, a finite representation (model) can be constructed for the covered part from the environment. Graph-based models and Petri net models to describe the discretized environment will be presented in Chapter 4. This chapter describes only cell decomposition approaches.

For mobile robots, cell decompositions are often employed in solving navigation problems, where a target position has to be reached without colliding with obstacles during movement. By using a cell decomposition of the free configuration space of the robot, one can solve a navigation problem by a graph search on the finite representation yielded by the decomposition [35, 92, 132]. Moreover, the involved model can provide a feasible way of solving coverage problems, where a robot should traverse all possible configurations without hitting obstacles. Cell decompositions were successfully used in problems where the task is given as a temporal and logic specification over some regions of interest from the environment [65, 111, 174]. In such scenarios, abstractions are usually obtained by decomposing the workspace of the system rather than its configuration space. More aspects on the general idea of cell decompositions can be found in [35].

Path Planning of Cooperative Mobile Robots Using Discrete Event Models, First Edition.
Cristian Mahulea, Marius Kloetzer, and Ramón González.

Despite the frequent use of cell decomposition techniques, there is a lack of user-friendly software frameworks performing multiple types of cell decompositions. Also, such frameworks would allow us to perform comparisons of different decomposition types, in order to help a user who is unsure which cell decomposition he/she should choose for trying to solve a specific problem.

This chapter has two main purposes. First, it describes the cell decompositions that include RMTool software from Chapter 2. Second, it provides qualitative and quantitative comparisons between the different decompositions. The decomposition routines start from the algorithms we reported in [113], which are presented in Section 3.2. We focus on two-dimensional environments cluttered with polygonal obstacles and on four types of decompositions, namely trapezoidal, triangular, polytopal, and rectangular. Section 3.3 gives further details on our implementation, provides additional examples, and illustrates the possibility of decomposing the defined obstacles. Based on the developed implementation, we perform extensive tests for comparing the decompositions and these results are given in Section 3.4.

3.2 Cell decomposition algorithms

3.2.1 Hypothesis

We consider a two-dimensional (2D) rectangular environment $Ev \subset \mathbb{R}^2$. Our restriction to two-dimensional environments is motivated by feasibility of the presented decompositions, and by the fact that many robotic problems consider planar environments [41, 191]. Specifically, the user defines the Cartesian coordinates $x_{min}, x_{max}, y_{min}, y_{max} \in \mathbb{R}$, and we have $Ev = \left[x_{min}, x_{max}\right] \times \left[y_{min}, y_{max}\right]$.

The planar environment is cluttered with n polytopal[1] obstacles (or regions), denoted by $O_1, O_2, \ldots, O_n$. We denote by $V_S = \{v_S^1 \ldots, v_S^{|V_S|}\}$ the set of vertices and by $F_S = \{f_S^1, \ldots, f_S^{|V_S|}\}$ the set of facets of any shape $S \in \{Ev, O_1, O_2, \ldots, O_n\}$. We use the convention of numbering the vertices and the facets of each region $S \in \{Ev, O_1, O_2, \ldots, O_n\}$ in a counter-clockwise order, by considering f_S^1 as defined by vertices v_S^1 and v_S^2.

Besides Ev, the user defines the Cartesian coordinates of vertices of each region, i.e. the set V_S, $\forall S \in \{O_1, O_2, \ldots, O_n\}$; see Figure 3.1. We mention that a polytope S can be mathematically represented (defined) in either of the two ways:

- the convex hull of its vertices (called V-representation or vertex-representation);

1 Through a "polytope" - a term inherited from arbitrary space dimensions - we understand the convex hull of a set of points (a bounded and convex shape with "flat" sides). In 2D, a polytope is a convex polygon.

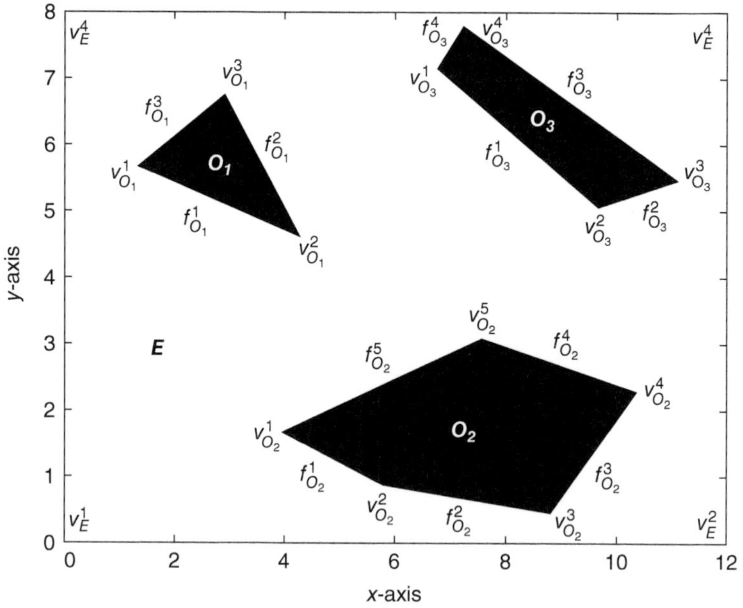

Figure 3.1 Example of an environment. An environment with three obstacles and the labeling of vertices and facets.

- the intersection of a number of half-spaces, each half-space being defined by a linear inequality (H-representation of face-representation). Each facet of S belongs to a (hyper-)plane defining a half-space.

The two types of representations can be interchangeably used, and there are algorithms for converting one representation to the other, for any space dimension [72, 157]. Thus, having V_S we can find F_S.

For simplicity of exposition of the decomposition algorithms, in this section we make the following assumptions:

A1 The defined regions are disjoint, i.e. $O_i \cap O_j = \emptyset$, $\forall i, j \in \{1, 2, \ldots, n\}$, $i \neq j$. Some cell decomposition algorithms from this section can be tailored to accommodate the case of overlapping regions, as will be noted in Section 3.3.

A2 The defined regions are considered as obstacles, i.e. only the free space (the subset of Ev not covered by any obstacle) will be partitioned into cells. Such a decomposition is suitable for obstacle avoidance (as a standard start to the goal reachability requirement in Chapter 2). For other mission types (LTL or Boolean formulas as in Chapter 2), we have regions of interest instead of obstacles (e.g. some cells should be visited or avoided, depending on the formula). In such cases, the regions of interest should also be captured by the cell decomposition. Again, two cell decomposition algorithms from this section can be tailored to partition regions, as will be noted in Section 3.3. Until then, we simply refer to defined regions as "obstacles" and we are interested in decomposing the free space.

Note that the assumption of having convex polygonal obstacles is not restrictive from a representation point of view, since any convex non-polygonal shape can be bounded by a convex polygonal region with arbitrary accuracy and any non-convex region can be divided into adjacent convex regions. The assumption of having a rectangular environment is common when representing the workspace in a robotic scenario, since the environment can be captured by an overhead video camera. Of course, the situation of a general polytopal environment can be achieved by appropriately defining inner obstacles near the edges of the rectangular environmental bounds.

The studied cell decomposition algorithms have as inputs $Ev, O_1, \dots, O_n$ (represented through their vertices). The outputs consist of a set of cells $C = \{c_1, c_2, \dots, c_{|C|}\}$ and an adjacency relation between cells, in the form of a symmetric matrix $A \in \mathbb{R}^{|C|\times|C|}$, where $A[i,j] = 1$ if cells c_i and c_j are adjacent and $A[i,j] = 0$ otherwise. We call adjacent cells two cells that share an entire line segment, not a single point. The intersection of any two adjacent cells has zero area and the union of all obtained cells is equal or close to the free space. All decompositions presented in this chapter are implemented in RMTool and indications for using the corresponding routines are given in Section 3.3.

This section presents four types of decomposition, the name of each of them being given by the shape of its cells:

- Subsection 3.2.2: "Trapezoidal" decomposition, yielding trapezoids with parallel vertical facets, while some trapezoids can become triangles.
- Subsection 3.2.3: "Triangular" decomposition, yielding triangles with vertices belonging to the set of algorithm inputs.
- Subsection 3.2.4: "Polytopal" decomposition, yielding convex poligons that have arbitrary (different) number of vertices.
- Subsection 3.2.5: "Rectangular" decomposition, yielding rectangles of different sizes, but with the same ratio width/height as environment Ev.

Cell decompositions in Subsections 3.2.2, 3.2.3 and 3.2.4 are also called "exact" decompositions, since the union of all obtained cells is equal to the free space. In contrast, the decomposition in Subsection 3.2.5 is called "approximate", since the union of the obtained cells is included in the free space. We do not consider other decomposition types, such as Boustrophedon or Morse [35], which have non-convex cells and are mainly used for specific coverage problems. Various advantages and disadvantages of the studied decompositions will be discussed in Section 3.4.

3.2.2 Trapezoidal decomposition

The trapezoidal decomposition [35] (sometimes called "vertical cell decomposition" [31, 135]) is often encountered with illustrative purposes in mobile robots problems. Its outcome is a partition of the free space with trapezoidal cells, where each trapezoid has vertical parallel edges. We note that using a similar idea, one can use a trapezoidal decomposition with cells having horizontal parallel sides.

The idea for constructing such a partition is to extend a vertical line from each vertex $v \in \bigcup_{i=1}^{n} V_{O_i}$, until facets from $\bigcup_{i=1}^{n} F_{O_i} \cup F_{Ev}$ are hit. The direction for extending each vertical line is given by the relative position of the current vertex with respect to other vertices from the same obstacle. For example, in Figure 3.1, the vertical line from vertex $v_{O_2}^2$ is extended only downwards, whereas the line from vertex $v_{O_2}^1$ is extended both upwards and downwards. When the line is extended in one direction, it will intersect one facet, while when it extends in both directions it will intersect two facets from $\bigcup_{i=1}^{n} F_{O_i} \cup F_{Ev}$. Once the line intersects these facets, it defines one or two new trapezoids to its left. For example, the line extended from $v_{O_1}^2$ will define two new trapezoids - cells c_4 and c_5 from Figure 3.2.

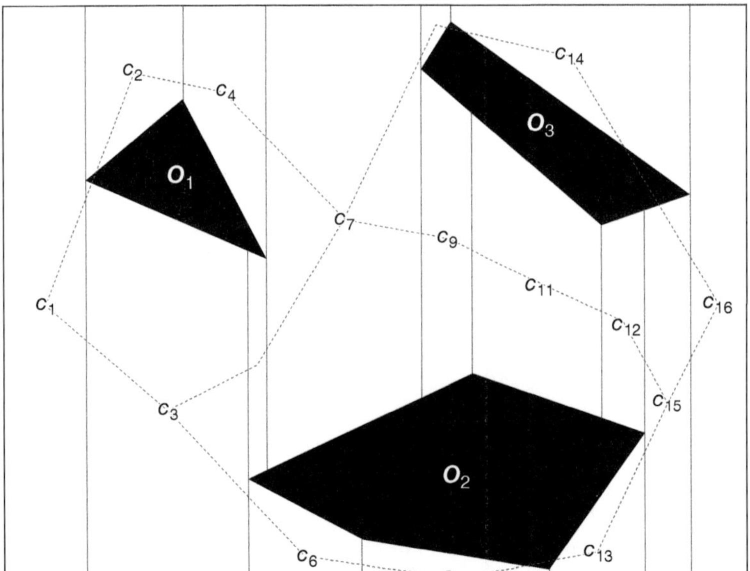

Figure 3.2 Trapezoidal decomposition example. Trapezoidal decomposition (consisting of 16 cells) of the environment from Figure 3.1. The adjacency relation is suggested by dotted lines connecting the centroids of adjacent cells. Note that the intersection of any two adjacent cells is a facet of one cell, but only a part of a facet for the other (this aspect that will be discussed in Subsection 3.4.1).

Algorithm 3.1: Trapezoidal decomposition

Input: $Ev, O_1, O_2, \dots, O_n$
Output: C, A

$C = \emptyset$
Let $\mathcal{V} = \bigcup_{i=1}^{n} V_{O_i}$ **while** $\exists v_a, v_b \in \mathcal{V}$ *with* $[0,1] \cdot v_a = [0,1] \cdot v_b$ **do**
  Perturb x-coordinate of v_b with a small amount
Sort vertices from $\mathcal{V}$ based on ascending order of their x-coordinate
while $\mathcal{V} \neq \emptyset$ **do**
  Choose the first vertex v from $\mathcal{V}$
  Extend a vertical line from v in the free space until it intersects one or two facets from $\bigcup_{i=1}^{n} F_{O_i} \cup F_{Ev}$
  Add the new trapezoid(s) isolated to the left of the extended line to C
  Remove v from $\mathcal{V}$
$A = \mathbf{0}_{|C| \times |C|}$
for $c_i, c_j \in C$, $c_i \neq c_j$ **do**
  if trapezoids c_i and c_j share a facet **then**
    $A[i,j] = 1, A[j,i] = 1$

Algorithm 3.1 presents an easy to follow pseudo-code of the corresponding procedure, which is implemented in RMTool. All obstacle vertices are first sorted based on their x-coordinate and then a vertical sweeping line is moved through each element of the sorted set of vertices. For further lowering the algorithm complexity, a list containing the facets intersected by the sweep line can be maintained and updated after handling each vertex in a specific manner. A detailed pseudo-code of that method can be found in [15, 35].

We note that, during trapezoidal decomposition, it is assumed that there are no obstacle vertices having the same x-coordinate [35, 135]. As in Algorithm 3.1, the situations not satisfying this assumption are handled by small random perturbations. Once the decomposition is performed, the adjacency relation is easily constructed by searching the pairs of cells that share a vertical line segment.

For illustrating the outcome of the algorithm, Figure 3.2 presents a trapezoidal decomposition of the environment from Figure 3.1.

3.2.3 Triangular decomposition

Let us first distinguish between the widely used Delaunay triangulations [70] and the triangular partitions we are interested in. A Delaunay triangulation receives as input a set of points (rather than obstacles) and connects these points in order

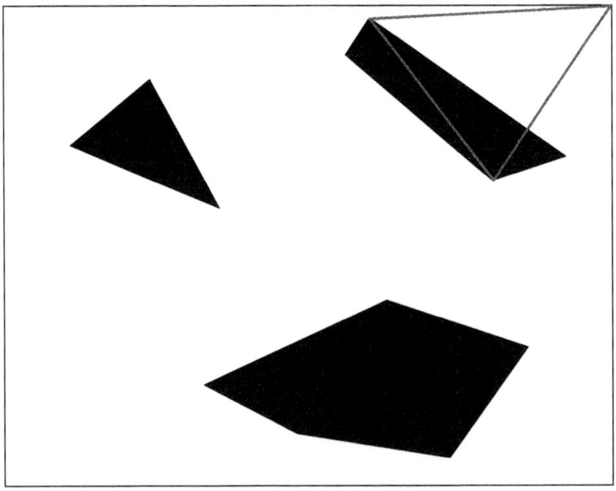

Figure 3.3 Problems with the standard Delaunay decomposition. Attempting to use a standard Delaunay decomposition (feeding to the algorithm only the vertices from Figure 3.1), would result in triangles as the red one, which violate the purpose of cell decompositions.

to form triangles with a special property - maximizing the minimum angle of all angles of the obtained triangles and ensuring that the circumscribed circle of each triangle does not contain any other point from the input set. A Delaunay triangulation cannot be used in our situation, because some of the obtained triangles may cut the obstacles, lying partially in the free space and partially in obstacles. For example, see in Figure 3.3 a triangle that would result from the Delaunay decomposition performed on the vertices defined in the environment from Figure 3.1.

To overcome this problem, constrained Delaunay triangulations were developed. They generalize standard Delaunay triangulation by guaranteeing that some predefined line segments are preserved [33, 190]. Some constrained Delaunay triangulation algorithms add additional points in order to maximize the minimum angle of triangles [190]. It is worth noting that algorithms for constrained Delaunay triangulations can run only in 2D environments, while standard Delaunay triangulations (called tessellations in higher dimension spaces) for points can be performed in any dimension [70, 190]. For our scenario, we focus on a constrained triangulation where no additional points are added. The predefined line segments are given by the set of obstacle facets, i.e. $\bigcup_{i=1}^{n} F_{O_i}$. Such triangulations are sometimes referred as triangulation of polygons with holes. Some algorithmic ideas for such procedures can be found in [15].

In our RMTool implementation, we adapt for our scenario some existing routines for constrained triangulation available in [1, 196]. Therefore, we cannot include specific pseudo-codes for these already available implementations. However, we mention that we use the Matlab's function "DelaunayTri". For 2D triangulations, this function receives two input arguments, a matrix X with two columns, each row containing the Cartesian coordinates of a point (vertex) in the environment, and a matrix *Cstr* with 2 rows, which defines the constrained edges, a row in this matrix containing the row indices of every two vertices from X defining an edge. Algorithm 3.2 includes the steps for interfacing

Algorithm 3.2: Triangular decomposition - interface with Matlab's function

Input: $Ev, O_1, O_2, \dots, O_n$
Output: C, A

$C = \emptyset$
$X = [x_{min}, y_{min}; x_{max}, y_{min}; x_{max}, y_{max}; x_{min}, y_{max}]$ (environment vertices)
$Cstr = [1, 2; 2, 3; 3, 4; 4, 1]$ (constraints given by indices of environment vertices)
for $i = 1, 2, \dots, n$ **do**
    for $v \in V_{O_i}$ **do**
        $X := [X; v^T]$ (add vertex to X)
for $i = 1, 2, \dots, n$ **do**
    for $f \in F_{O_i}$ **do**
        Find j and k such that facet f defined by vertices on rows j and k of X
        $Cstr := [Cstr; j, k]$ (add constraint for facet f)
Call Matlab function *Triang = DelaunayTri(X, Cstr)*
Triang.Triangulation contains indices from X defining triangles
for any row i of *Triang.Triangulation* **do**
    //Any returned triangle is either fully inside an obstacle or fully in free space
    if centroid of current triangle is not contained in any obstacle **then**
        Add current triangle to C
$A = \mathbf{0}_{|C| \times |C|}$
for $c_i, c_j \in C$, $c_i \neq c_j$ **do**
    if triangles c_i and c_j share two vertices **then**
        $A[i, j] = 1, A[j, i] = 1$

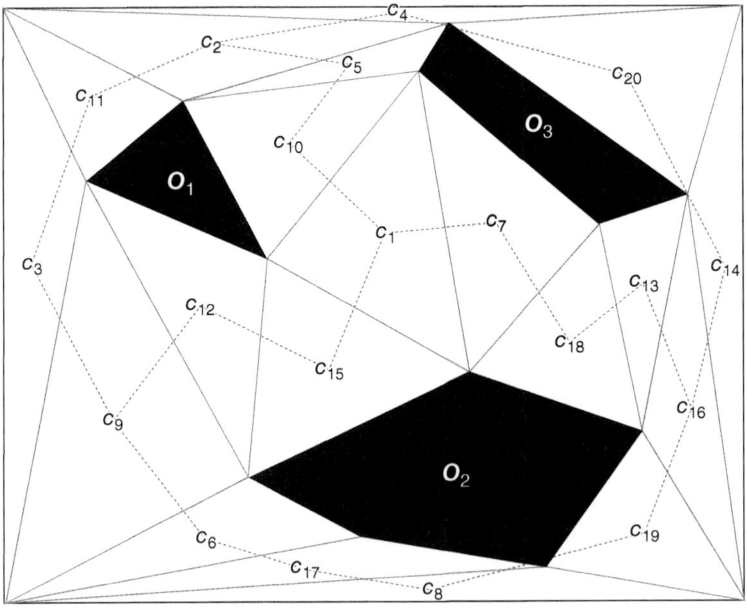

Figure 3.4 Constraint triangular decomposition. Constrained triangular decomposition of the environment from Figure 3.1. Here 20 cells are obtained, and the adjacency relation is represented by dotted lines connecting the centroids of adjacent cells. Triangular decompositions can be tailored to also partition the defined regions (obstacles), as will be mentioned in Subsection 3.3.1.

our scenario with the mentioned Matlab function. A resulting constrained triangular decomposition of the environment from Figure 3.1 is presented in Figure 3.4.

3.2.4 Polytopal decomposition

A polytopal decomposition consists of polytopal cells having a varying number of vertices. The idea for constructing such a decomposition is to extend the supporting line for each obstacle facet and then to search all the polytopes defined by such lines in *Ev* [111]. Each obtained cell is defined by a set of lines, and no other facet-supporting line should intersect that cell.

Algorithm 3.3 presents the main steps for creating a polytopal decomposition. First, the equations of each line supporting a facet are found (as lines defined by each two adjacent vertices of the convex hull of a region) and added in matrix P (having two columns) and column vector Q (lines 2-7 of Algorithm 3.3). Each such line defines two half-spaces, and Algorithm 3.3 tests every possible

Algorithm 3.3: Polytopal decomposition

Input: $Ev, O_1, O_2, \dots, O_n$
Output: C, A

1 $C = \emptyset$
2 Let P be an empty matrix and Q an empty vector
3 **for** $S \in \{Ev, O_1, O_2, \dots, O_n\}$ **do**
4   **for** $f \in F_S$ **do**
5   Construct supporting line of facet f, defined by $p, q: f \in \{X \in \mathbb{R}^2 | pX + q = 0\}$
6   Add line p to matrix P
7   Add element q to vector Q
8 **for** $i \in \{0, 1, \dots, 2^{|Q|-1}\}$ **do**
9   Let $b_0 b_1 \dots b_{2^{|Q|-1}}$ be the binary representation of i
10     $D = \mathbf{0}_{2^{|Q|-1} \times 2^{|Q|-1}}$
11     $D[j, j] = -1$ if $b_j = 0$, $D[j, j] = 1$ if $b_j = 1$, $\forall j = 0, 1, \dots, 2^{|Q|-1}$
12   Let $Polytope = \{X \in \mathbb{R}^2 | D \cdot P \cdot X \leq D \cdot Q\}$
13   **if** $Polytope \neq \emptyset$ **then**
14   Find set $V_{Polytope}$ containing vertices of *Polytope*
15   **if** $(V_{Polytope} \subset Ev)$ **and** $(V_{Polytope} \setminus O_j \neq \emptyset,\ \forall j = 1, \dots, n)$ **then**
16   Add $V_{Polytope}$ to set of cells C
17 $A = \mathbf{0}_{|C| \times |C|}$
18 **for** $c_i, c_j \in C$, $c_i \neq c_j$ **do**
19   **if** $|V_{c_i} \cap V_{c_j}| = 2$ **then**
20         $A[i, j] = 1, A[j, i] = 1$

non-empty intersection of all such half-spaces (lines 8-16). Note that checking the emptiness of a polytope defined by a set of linear inequalities (*Polytope* on line 12 of Algorithm 3.3) can be either performed by using specific routines for polytopes, e.g. [72], or can be accomplished by solving a linear programming problem having the same linear inequalities as the constrained set. If the polytope is non-empty, we find its vertices (line 14). We mention that, whenever a set of linear inequalities defines a non-empty polytope, the vertices are constructed by converting the face-representation (given by $D \cdot P \cdot X \leq D \cdot Q$ in Algorithm 3.3) to a vertex-representation [72, 157]. If the found polytope lies in the free space, it is added to the cell set. Because of the construction we use, a non-empty polytope either (i) lies in the free space or (ii) is outside environment *Ev* or (iii) all its vertices are contained by a single obstacle. The test from line 15 of Algorithm 3.3 checks the negation of cases (ii) and (iii) for deciding if case (i) is fulfilled.

After all feasible cells are found, the adjacency matrix A is constructed by finding pairs of cells that share two vertices (lines 17-20 of Algorithm 3.3). This simple adjacency construction can be performed because our partition has the property that any two adjacent polytopes exactly share a facet.

For simplicity of exposition, the construction in Algorithm 3.3 contains a brute-force approach for finding all feasible intersections of half-spaces. In our implementation, we lower this complexity by using (before loop in line 8) an iterative procedure that starts with just a few elements from P and Q, finds their feasible intersections, and then adds more elements. This procedure basically restricts at each step the set of values of some bits $b_0, b_1, \dots, b_{2|Q|-1}$ (e.g. including the positions corresponding to Ev, which should always have fixed signs). Thus, we end up with testing a number of combinations much smaller than $2^{|Q|-1}$.

A polytopal decomposition of the environment from Figure 3.1 is given in Figure 3.5.

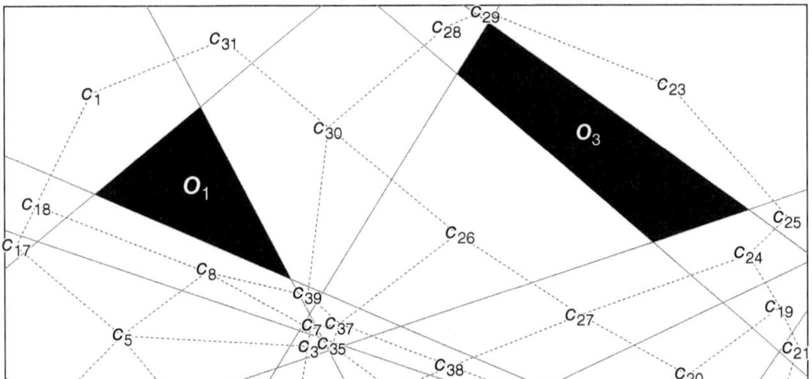

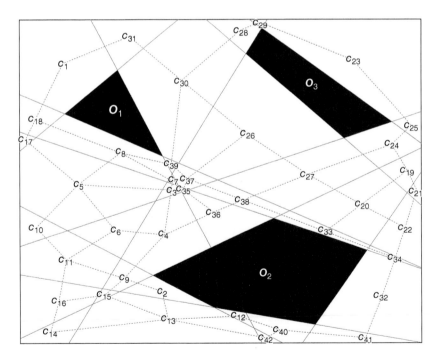

Figure 3.5 Polytopal decomposition. Polytopal decomposition consisting of 42 cells of the environment from Figure 3.1. Dotted lines connect the centroids of adjacent cells. Usually, polytopal decompositions produce many small cells, an aspect that will be detailed in Section 3.4.

3.2.5 Rectangular decomposition

As mentioned, the rectangular decomposition is an approximate cell decomposition, that covers the free space up to an arbitrary precision. This decomposition is inspired by quad-trees [15, 68], which are hierarchical decompositions mainly used in computer science for organizing data or compressing images. A quad-tree is a tree whose nodes are either leaves (have no children) or have four children. Each node in a quad-tree corresponds to a rectangle in the environment Ev and is labeled with one of the three symbols: *occupied* - if all points in the corresponding rectangle are included in one obstacle or a union of obstacles, *free* - if no points in the rectangle are contained in any obstacle, or *mixed* - if the rectangle contains both points which are in obstacles and points that are not in obstacles. The main idea of this decomposition is to recursively split every mixed rectangle (parent) in four equal subrectangles (children) by dividing each of its sides in to two equal parts. A rectangle is not split any more if it is free, occupied, or mixed with dimensions smaller than a given precision ϵ. In the last situation, the rectangle is labeled as occupied. Thus, in the end, the labels of all leaves are occupied or free. The cells of the decomposition consist of all free leaves, and the adjacency relation is constructed based on their relative position in Ev.

The procedure we develop for recursively labeling and splitting mixed rectangles is given by Algorithm 3.4. For performing the rectangular decomposition, this procedure is called from a main program with arguments $C = \emptyset$, limits of environment Ev, and a desired precision ϵ. The labeling of the current rectangle R from Algorithm 3.4 is decided based on the area of R that overlaps with obstacles (computed in line 4). For computing this area, one can use tools that perform intersections of polyhedra [72]. If R is free, it is added to current decomposition C and the procedure finishes. Whenever R is mixed, it is split into four equal subrectangles and the procedure from Algorithm 3.4 is recursively run (lines 10-13). An occupied or too-small rectangle R does not influence the current decomposition C. Once C is found, the adjacency is constructed by finding pairs of rectangles that share a line segment.

Observe that, due to performed splitting, every rectangle from decomposition C will have the same width/height ratio as environment Ev. Also, precision ϵ can be defined based on a dimension of Ev and it thus limits the maximum number of splits. For example, if d denotes the smallest side of Ev (minimum from width and height), then by choosing any ϵ in interval $[d/8, d/4)$ will yield that the smallest cell has sides 4 times smaller than Ev.

A rectangular decomposition of the environment from Figure 3.1 is illustrated by Figure 3.6, where precision ϵ was chosen 32 times smaller than the dimension of Ev on he y-axis.

Algorithm 3.4: Procedure *check_split_rectangle*

Input: Current decomposition $C, x_{min}, x_{max}, y_{min}, y_{max}, \epsilon, O_1, O_2, \ldots, O_n$
Output: Updated decomposition C

$R = [x_{min}, x_{max}] \times [y_{min}, y_{max}]$ is the current rectangle
if $(x_{max} - x_{min} \leq \epsilon)$ **or** $(y_{max} - y_{min} \leq \epsilon)$ **then**
  return C /* R is too small */
$area_{intersect} = \bigcup_{i=1}^{n} Area(R \cap O_i)$
if $area_{intersect} = 0$ **then**
  Add R to C /* R is free */
  return C
else
  if $area_{intersect} < Area(R)$ /* R is mixed */ **then**
    $C = check\_split\_rectangle\left(C, x_{min}, \frac{x_{min}+x_{max}}{2}, y_{min}, \frac{y_{min}+y_{max}}{2}, \epsilon\right)$
    $C = check\_split\_rectangle\left(C, \frac{x_{min}+x_{max}}{2}, x_{max}, y_{min}, \frac{y_{min}+y_{max}}{2}, \epsilon\right)$
    $C = check\_split\_rectangle\left(C, \frac{x_{min}+x_{max}}{2}, x_{max}, \frac{y_{min}+y_{max}}{2}, y_{max}, \epsilon\right)$
    $C = check\_split\_rectangle\left(C, x_{min}, \frac{x_{min}+x_{max}}{2}, \frac{y_{min}+y_{max}}{2}, y_{max}, \epsilon\right)$
  else
    return C /* R is occupied */

Note that we do not analyze the direct approach of splitting the environment into small rectangles with the same area, as a grid with constant increments on a 2D axis, and then deciding which of these rectangles are free. When employed, such a decomposition usually yields a large number of cells and it does not have an increased resolution around obstacles, as in the presented approach.

3.3 Implementation and extensions

The four cell decompositions from Section 3.2 are included in RMTool, together with routines for defining an environment as in Section 3.2.1. The implementation is carried out in MATLAB [196] and includes some functions from the libraries found in [72, 199] for performing polyhedral operations.

3.3.1 Extensions

As one can notice from Section 3.2, triangular and polytopal decompositions have the property that adjacent cells share facets exactly. Due to this reason,

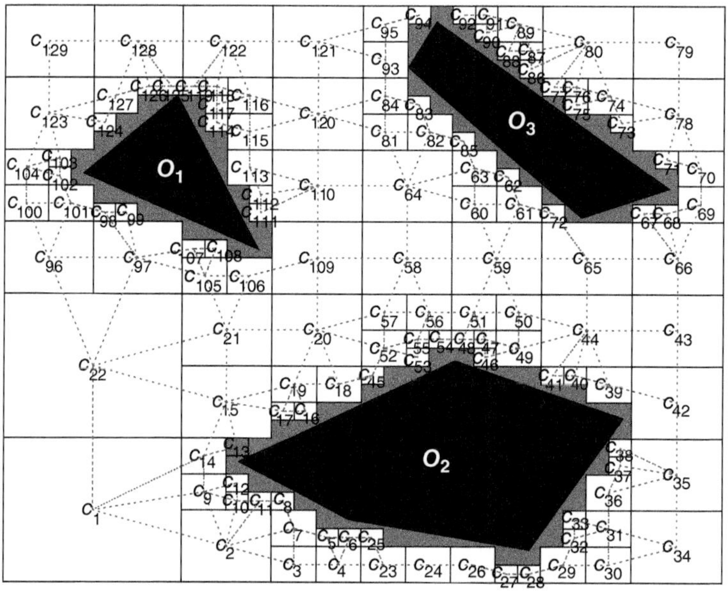

Figure 3.6 Rectangular decomposition. Rectangular decomposition (consisting of 129 cells) of the environment from Figure 3.1. The gray area around obstacles is considered as occupied, due to approximate cell decomposition. Dotted lines connect the centroids of adjacent cells. Although there are many small cells, these tend to appear only around obstacles.

they are frequently used for solving robotic problems where the task requires visiting some regions (obstacles from above), rather than avoiding all of them [65, 111]. In such cases, both the free space and the defined regions have to be decomposed into cells. Therefore, we extended our procedures for triangular and polytopal decompositions, such that a partition including the defined regions can be returned. In such a case, additional outputs result, showing the inclusion of each cell in the free space or in region $O_1, \ldots, O_n$. An example of a triangular decomposition where the defined regions are also decomposed is given in Figure 3.7.

Moreover, each decomposition procedure returns (if desired) the middle points of the line segments shared by adjacent cells, e.g. as in Figure 3.7. These middle points are useful in solving robotic navigation problems via cell decompositions, where a trajectory through cells can be mapped into a trajectory consisting of line segments linking such middle points [35, 135]; more details will be given in Chapter 5.

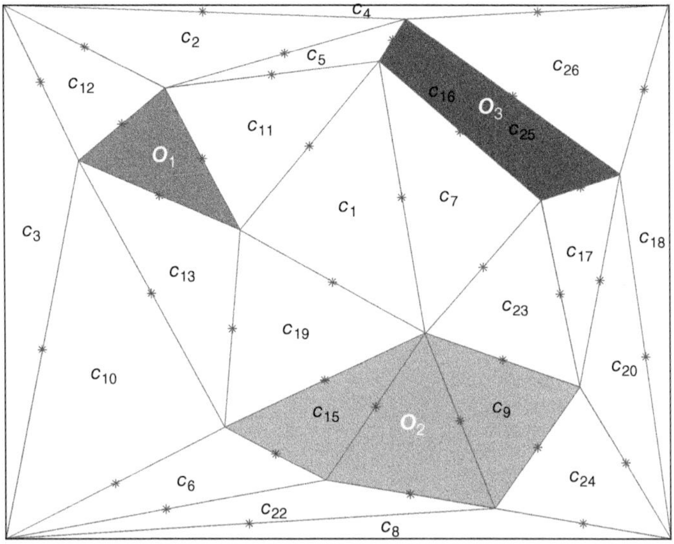

Figure 3.7 Triangular decomposition with regions of interest. Triangular partition consisting of 26 cells of the environment from Figure 3.1. The regions are also decomposed and the middle points of segments shared by adjacent cells are represented by red stars.

3.3.2 Implemented functions

The cell decomposition routines can also be run in Matlab, outside the provided RMTool's Graphical User Interface. For these functions to run properly, the auxiliary toolboxes of RMTool (folder "aux_toolboxes") should be in Matlab's path. This subsection explains the syntax and usage in Matlab of functions related to cell decompositions; the names of all these RMTool functions begin with "rmt".

Environment definition: function *rmt_define_regions*

The function allows the user to set the environment bounds and the number of obstacles, and then to choose obstacle vertices via mouse clicks.

Syntax: *Obs* = *rmt_define_regions*(*Ev_bounds*, *n*)

- *Ev_bounds* is a vector of form $[x_{min}, x_{max}, y_{min}, y_{max}]$, where $x_{min}, x_{max}, y_{min}, y_{\max} \in \mathbb{R}$ are the Cartesian coordinates of the rectangular environment *Ev*;
- *n* is the number of obstacles to define;
- *Obs* is a Matlab cell array containing obstacles. *Obs*$\{i\}$ is a matrix with 2 rows, where each column contains the Cartesian coordinates of O_i. Specifically, in Matlab, *Obs*$\{i\}(:,j)$ returns vertex $v^j_{O_i}$.

Note that the vertices of each obstacle are defined by mouse clicks: the left-click selects a vertex and the right-click selects the last vertex of the current obstacle;

then the function waits for vertices of the next region. If the user clicks outside the environment bounds, the point is not stored. Moreover, if a right-click is used before properly ending the obstacle (e.g. there are less than 3 vertices for the current region, or all vertices are collinear and thus they do not form a convex shape with non-zero area), the vertex is stored, but the reading for points of the current region continues.

For each obstacle O_i, no matter the order in which vertices are defined, columns of matrix $Obs\{i\}$ contain only the vertices defining the convex hull of O_i, and these vertices are stored in a counter-clockwise order. The redundant defined vertices (inside O_i) are removed.

Cell decomposition: functions *rmt_***type***_decomposition*, where **type** is to be replaced with either one of *trapezoidal, triangular, polytopal, or rectangular.*

The functions perform the desired cell decomposition of an environment. The input arguments of each procedure consist of the defined environment and obstacles, appended with precision ϵ in the case of rectangular decompositions. The outputs include cell decomposition, adjacency matrix, and middle points of segments shared by adjacent cells.

Syntax: $[C, A, mid\_x, mid\_y] = rmt\_\mathbf{type}\_decomposition(Obs, Ev\_bounds, \epsilon)$

- *Obs* is a cell array containing vertices of the *n* obstacles, e.g. as returned by function *rmt_define_regions*;
- *Ev_bounds* gives the environment bounds, as in function *rmt_define_regions*;
- ϵ is a number defining the precision, and it should be given only for rectangular decomposition;
- C is a cell array, where $C\{i\}$ is a matrix with 2 rows containing the coordinates of vertices of cell c_i from the decomposition. Command *length (C)* gives the number of obtained cells, i.e. $|C|$;
- A is the symmetric adjacency matrix, as defined in Section 3.2.1. Recall that $A[i,j] = 1$ if cells c_i and c_j are adjacent, and adjacent cells mean that they share a line segment (not just a point). For optimizing memory usage, A is stored as a sparse matrix in Matlab, since many of its elements are zero;
- *mid_x,mid_y* give the middle points of shared line segments. For any adjacent cells c_i and c_j, the middle point of the line segment shared by them has the x-coordinate *mid_x(i,j)* and the y-coordinate *mid_y(i,j)*. The pair of outputs *mid_x,mid_y* can be removed from the syntax if desired.

For example, the Matlab command "$[C, A] = rmt\_triangular\_decomposition$ $(\{[1, 3, 4; 2, 4, 1], [6, 9, 8; 6, 8, 5]\}, [0, 10, 0, 10])$" performs the triangular decomposition of an environment containing two triangular obstacles.

Cell decomposition with regions: functions *rmt_***type***_decomposition_regions*, where **type** is to be replaced with *triangular* or *polytopal.*

The defined shapes are regarded as regions of interest, not obstacles, and they are partitioned as well. Moreover, the defined shapes are allowed to overlap, since the user may want some points in the environment to belong to more than one region of interest.

In such a case, additional outputs (*Observ_set* and *observ*) show where each cell from the decomposition lies (in the free space or in region $O_1, \ldots, O_n$).

Syntax: $[C, A, Observ\_set, observ, mid\_x, mid\_y] = rmt\_\mathbf{type}\_decomposition\_regions(Reg, Ev\_bounds)$

- *Reg* has exactly the same structure as *Obs* from above, but the obstacles (regions) may now intersect;
- *Ev_bounds* gives the environment bounds, as above;
- *C* and *A* give the cells and the adjacency matrix, as above;
- *Observ_set* is a matrix with n columns, giving in rows the possible observations (region overlapping). The rows are padded with zeros up to n elements. A row such as $[1, 3, 0]$ shows that there are 3 defined obstacles, and there exists an overlapping of first and third, with observation 1 and 3. The last row of *Observ_set* corresponds to observing the free space, and has the form $[n + 1, 0, \ldots, 0]$, i.e. the $n + 1$ index corresponds to not observing any region. Basically, *Observ_set* is created such that it contains feasible elements of the power set $2^{\{O_1, O_2, \ldots, O_n\}}$;
- *observ* is a vector with $|C|$ elements, where *observ(i)* gives the index of observation of cell c_i, i.e. the index of row from *Observ_set*. For example, *Observ_set(observ(i),:)* shows all the defined regions that contain cell c_i. If *Observ_set(observ(i),:)* is $[1, 3, 0]$, then c_i belongs to the intersection of the first and third defined regions, whereas if *Observ_set(observ(i),:)* is $[n + 1, 0, 0]$, then c_i lies in the free space.
- *mid_x,mid_y* give the middle points of shared line segments. They have the same structure as above, and they are optional.

Graphical representation: function *rmt_plot_environment*

The function allows the user to represent an environment with obstacles, and if desired the cells resulted from decomposition, adjacency relations, and middle points.

Syntax:

- *rmt_plot_environment(Obs,Ev_bounds)* represents the defined obstacles, with *Obs* and *Ev_bounds* defined as above;
- *rmt_plot_environment(Obs,Ev_bounds,C)* additionally represents the cell decomposition given by *C*, without labeling (numbering) cells;
- *rmt_plot_environment(Obs,Ev_bounds,C,string)* represents the cell decomposition and labels cells in form "$string_i$", with $i = 1, \ldots, |C|$. For example, if *string* ='\pi', cells will be labeled with $\pi_1, \pi_2, \ldots$, while for the above figures *string* ='c'. Obstacles are labeled with $O_i, i = 1, \ldots, n$;

- *rmt_plot_environment(Obs,Ev_bounds,C,string,A)* also suggests adjacency, by dotted lines linking centroids of neighboring cells. Use '[]' instead of *string* for not displaying cell and obstacle labels;
- *rmt_plot_environment(Obs,Ev_bounds,C,string,A,mid_x,mid_y)* represents cells and marks the middle points of line segments shared by adjacent ones, e.g. as in Figure 3.7.

Figures 3.2, 3.4, 3.5, and 3.6 contain various decompositions of the environment from Figure 3.1, while Figures 3.8a to d present decompositions of a more complex environment. Representations in Figure 3.8 support a better understanding of some qualitative aspects of decompositions that will be given in Section 3.4.

3.4 Comparative analysis

This section performs a comparative study of the presented cell decompositions, with the goal of providing some recipes that can be used for choosing a certain decomposition over the others for a specific problem. We note that the present study relies on comparing the decompositions, and it does not include comparisons for planning robot trajectories. Such extended correlations will be given in Chapter 5.

Subsection 3.4.1 provides a qualitative comparison of the studied cell decompositions and Subsection 3.4.2 includes quantitative tendencies of different cost criteria, computed by using the implementations mentioned in Section 3.3.

3.4.1 Qualitative comparison

The qualitative aspects from this section can be useful for quickly ruling out some cell decomposition types that do not have a specific property required for a certain problem. This qualitative comparison includes four different aspects, which are detailed below.

i) *Exactly share facets*: In the triangular and polytopal decompositions, all the adjacent cells exactly share facets (meaning that the intersection of any two adjacent cells c_i and c_j is a facet for both c_i and c_j), while this is not true for the trapezoidal and rectangular decompositions. For example, in Figures 3.2 and 3.6 the intersection of adjacent cells c_1 and c_2 is an entire facet of c_2, but it is only a part of a facet of c_1. In contrast, the intersection of any two adjacent cells from Figures 3.4 and 3.5 is equal to an entire facet of both cells. This property is useful when one can design feedback continuous controllers in each cell such that the cell is left through a desired facet (exit-to-facet control), as in [90]. Leaving a cell through a desired facet guarantees what the

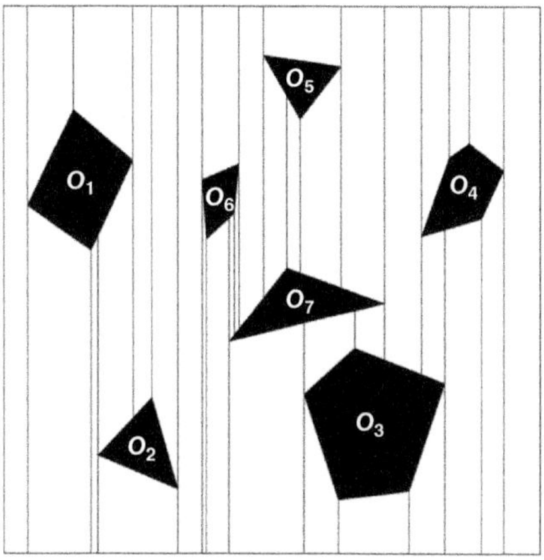

(a) Trapezoidal decomposition consisting of 35 cells.

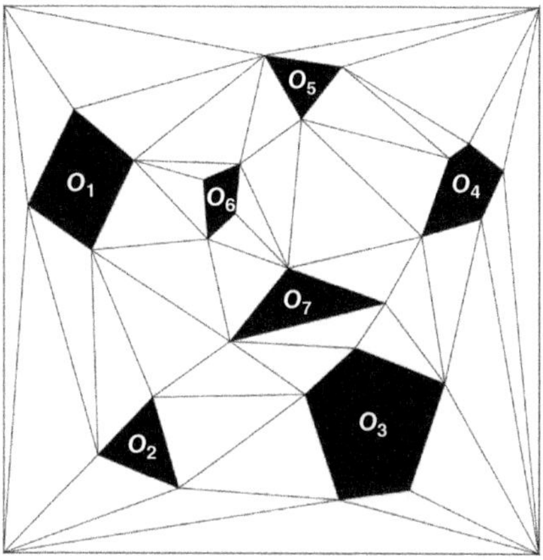

(b) Constrained triangular decomposition with 43 cells.

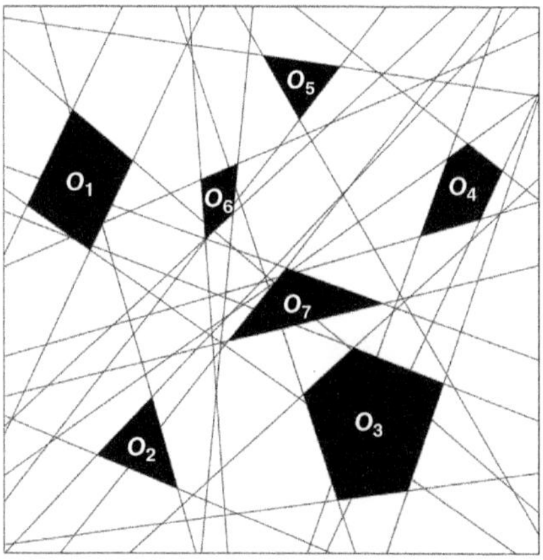

(c) Polytopal decomposition consisting of 220 cells.

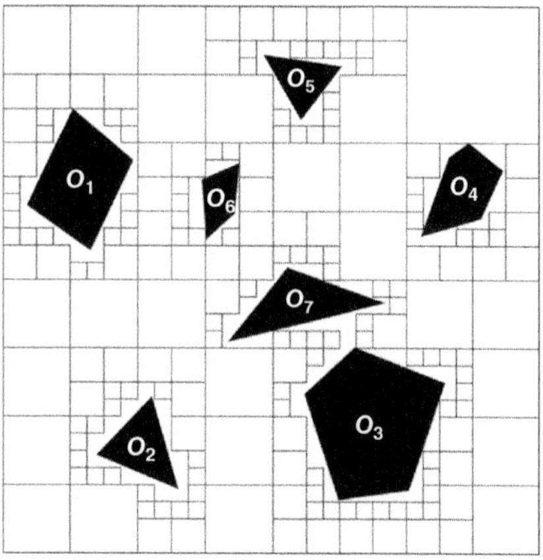

(d) Rectangular decomposition with 194 cells. For the rectangular decomposition, the sides of the smallest cell were 32 times smaller than the sides of the environment.

Figure 3.8 Comparison of cell decomposition approaches. Cell decompositions for an environment containing seven obstacles.

next visited cell is, whereas this result obviously does not hold for decompositions where adjacent cells do not exactly share facets. Thus, when using a polytopal or triangular decomposition, one can easily map a desired sequence of cells to be followed by a mobile robot with affine dynamics to a sequence of affine controllers designed as in [90]; such an approach was successfully used in [111]. In contrast, when a sequence of cells should be followed in a trapezoidal or rectangular decomposition, a simpler dynamics can be considered for a mobile robot (usually fully actuated) and an angular path can be followed, as suggested in [35]. If one wants to use rectangular or trapezoidal partitions with exit-to-facet controllers, a subpartitioning of the traversed cells may be involved, as in [110].

ii) *Higher dimension spaces*: In space dimensions higher than two, the polytopal and rectangular decompositions are straightforward to generalize, although with a corresponding increase in computational needs. The trapezoidal and triangular decompositions can be generalized only under specific restrictions [135, 190], and the algorithms would become much harder to develop than in the two-dimensional case. Therefore, if one should use a cell decomposition in space dimensions higher than two, we suggest a generalization of either the polytopal or the rectangular decomposition algorithms from Subsections 3.2.4 and 3.2.5. For example, such a generalization of rectangular decompositions is used in [110] for planning a robot in a 3D environment.

iii) *Scattered small cells*: The trapezoidal, polytopal, and rectangular decompositions tend to have a large number of cells with a small area. For the trapezoidal and polytopal decompositions, such cells generally appear in various places in the environment, and it may be difficult or impossible to find a path that avoids them. Moreover, in the case of trapezoidal decompositions, such cells result from small perturbations that are needed when more vertices have the same x-coordinate (see Subsection 3.2.2). In contrast, the rectangular decomposition has increased resolution (more small cells) only around obstacles. To support these statements, we refer to the decomposition examples in Figure 3.8 and in Section 3.2. Note that cells with a small area may be disadvantageous when controlling a mobile robot with a feedback control law based on its current position. This is because small errors in knowing the actual robot position may wrongfully report that the robot is in another cell, while errors in moving the robot may result in leaving the desired cells and even hitting obstacles. Therefore, in situations when there is not very good precision in acquiring the robot position and in controlling its motion, we suggest avoiding trapezoidal and polytopal decompositions. When using a rectangular decomposition, evolving through cells with a large area (e.g. by imposing a penalty for visiting small cells) usually keeps the mobile robot far from obstacles. Additional quantitative results on cells with a small area will be included in Subsection 3.4.2.

Table 3.1 Properties (on rows) versus cell decomposition types (on columns).

	Trapezoidal	Triangular	Polytopal	Rectangular
Exact decomposition	Yes	Yes	Yes	No
Adjacent cells exactly share a facet	No	Yes	Yes	No
Straightforward to generalize to higher space dimensions	No	No	Yes	Yes
Many scattered small cells are obtained	Yes	No	Yes	No
Greatly affected by small environment modifications	No	No	Yes	No
Allows the partition of regions of interest	No	Yes	Yes	No

iv) *Imprecise environment map*: Due to its construction, the polytopal decomposition is affected by small modifications in the environment. For example, if a vertex of an obstacle is slightly moved, the supporting lines containing that vertex suffer changes that might affect many cells (new cells may appear and some other cells may not be feasible polytopes any more). This aspect may rule out the polytopal decomposition when a noisy environment map is available. A cost criterion corresponding to this aspect will be included in Subsection 3.4.2.

v) *Regions of interest*: As mentioned in Section 3.3, the polytopal and triangular decompositions were extended in our implementation such that the defined regions are also partitioned. Note that in such cases, we do not talk about "obstacles"; we rather use the term "region" or "region of interest", because partitioning these regions is meaningful if (some of them) should be visited, not avoided. Indeed, such scenarios are used when planning a mobile robot based on more complex requirements than just avoiding obstacles, as will be presented in subsequent chapters. Note that even when the regions are decomposed, the polytopal and triangular decompositions maintain the property of producing adjacent cells that entirely share a facet.

Table 3.1 summarizes the discussed properties of the four types of cell decompositions.

3.4.2 Quantitative comparison

The qualitative aspects in Subsection 3.4.1 can yield multiple feasible decomposition choices for a certain scenario. The comparison in this section gives a quantitative information on several aspects related to each decomposition. This data is embedded in a number of cost criteria, each being dependent on the number of obstacles in the environment. For measuring the cost criterion values, we considered a number of obstacles n ranging from 1 to 10, and for each choice of n we randomly created 1000 environments (with different position, size, and number of vertices for each obstacle). Each cell decomposition algorithm was run on these environments. The average values of the 1000 trials were computed for each number of obstacles, for each cost criterion, and for each decomposition.

In the following, we present the considered cost criteria (metrics) and we interpret the obtained evolution versus the number of obstacles:

i) The total number of cells, $|C|$ (Figure 3.9(a)): the depicted tendency shows that the number of cells yielded by the polytopal decomposition abruptly increases with the number of obstacles, when compared to other decompositions.
ii) The computation time (Figure 3.9(b)) necessary for performing a polytopal decomposition is more disadvantageous than for other decomposition types; this fact is supported by the large number of tests for feasible polytopes from Algorithm 3.3. The computation times depicted in Figure 3.9(b) were obtained by performing cell decompositions on a standard-performance computer (2GHz processor, 4GB RAM). Note that these small computation times support the possible usefulness of our implementations for real experiments.
iii) The number of adjacency relations (Figure 3.9(c)) has a similar increasing trend as criteria (i) (Figure 3.9(a)). The cost criteria (i) and (iii) are a measure of the complexity for performing a search in the discrete event model corresponding to a decomposition. We did not include the number of vertices of resulting cells (which affects the storage requirements), since it has a very similar shape to the one in Figure 3.9(c).
iv) Average area of cells from decomposition (Figure 3.9(d)): the choice for this metrics is mainly motivated by the applicability of cell decompositions in robot motion planning, where it is desirable to move a robot through larger cells in order to gain robustness with respect to closeness to obstacles and to noise in acquiring the exact position of the robot. The illustrated tendencies show that the trapezoidal and triangular decompositions seem more promising, while the polytopal decomposition is disadvantageous, especially for environments cluttered with many obstacles. The results in Figure 3.9(d) referring to the rectangular decomposition should be regarded with caution, because they are negatively altered by aspects mentioned in qualitative criterion (iii) in Subsection 3.4.1.
v) The standard deviation of cell area was recorded for each experiment and the average values over the 1000 environments are depicted in Figure 3.9(e). This metrics is included as a measure of diversity in cell area (cost (iv)) and suggests that the trapezoidal and triangular decomposition are affected by the

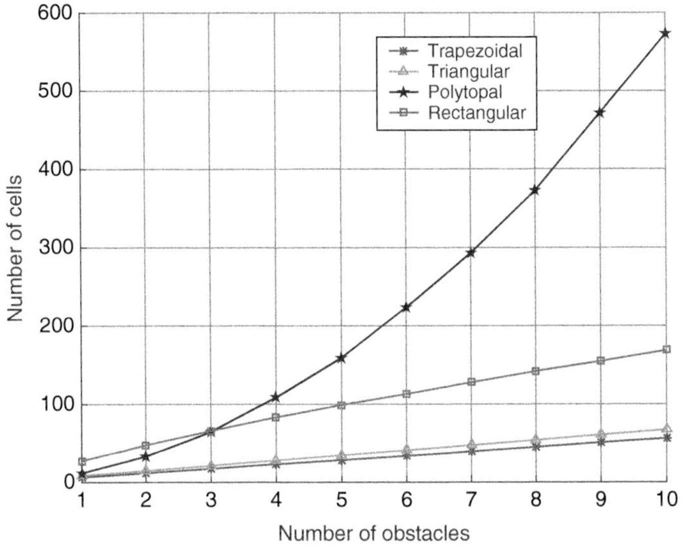

(a) Number of cells.

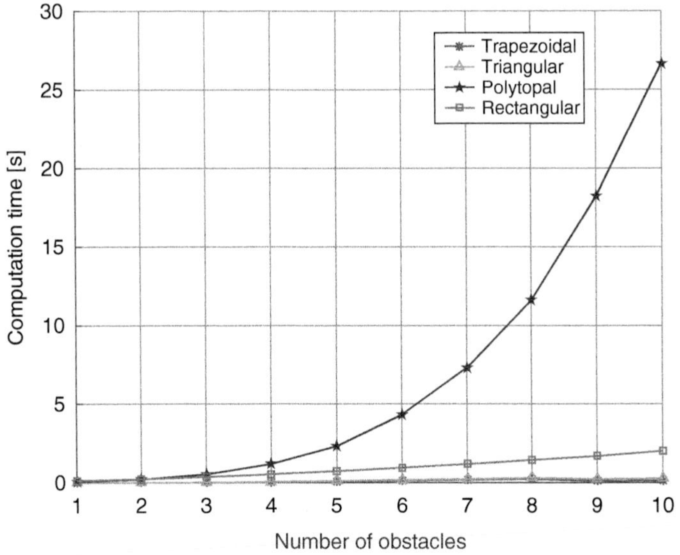

(b) Computation time (in seconds).

Figure 3.9 Comparison of cell decomposition approaches. Criteria for comparing cell decompositions, obtained by performing tests over 1000 randomly generated environments. All metrics are presented versus the number of obstacles in the environment.

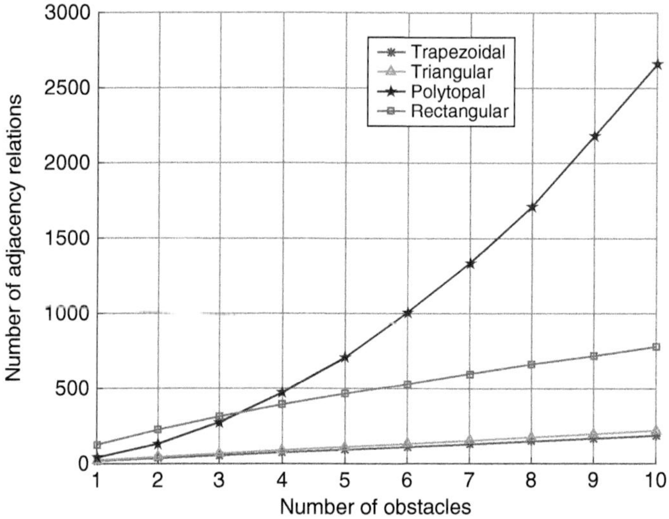

(c) Number of adjacency relations.

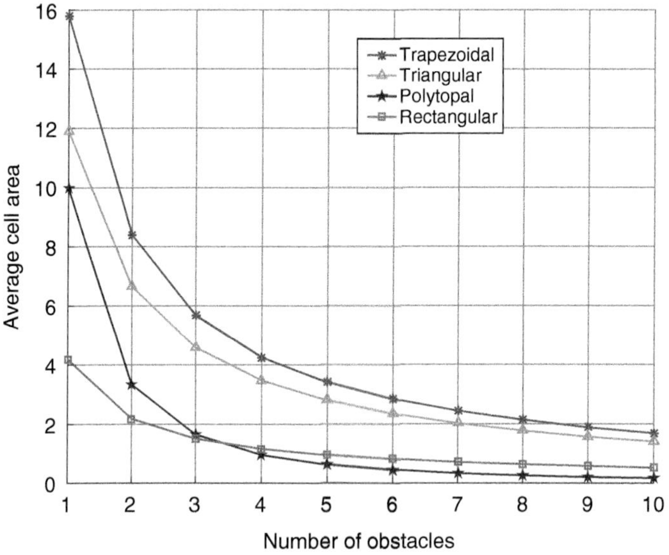

(d) Average area of cells.

Figure 3.9 (*Continued*)

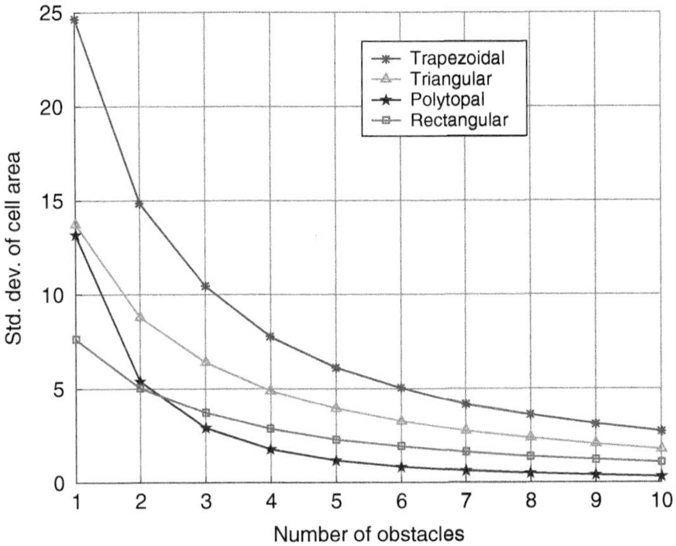

(e) Standard deviation of cell area.

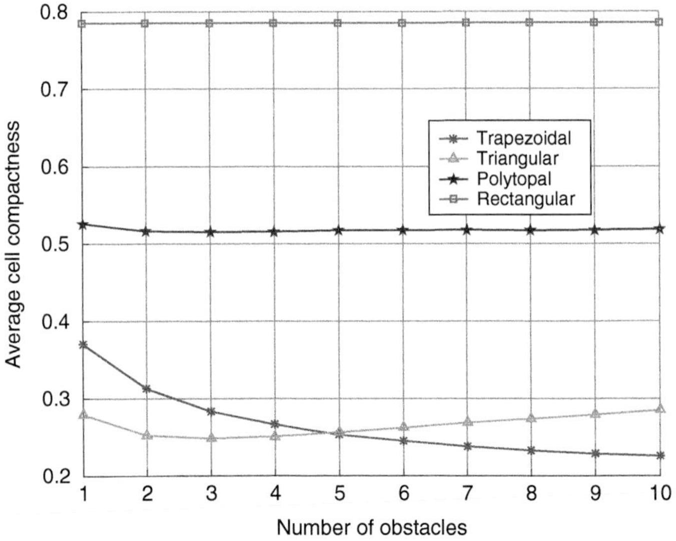

(f) Average compactness of cells.

Figure 3.9 *(Continued)*

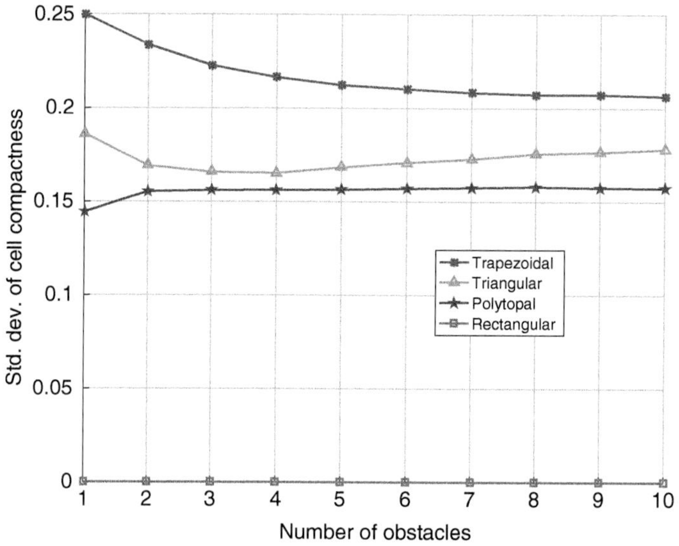

(g) Standard deviation of cell compactness.

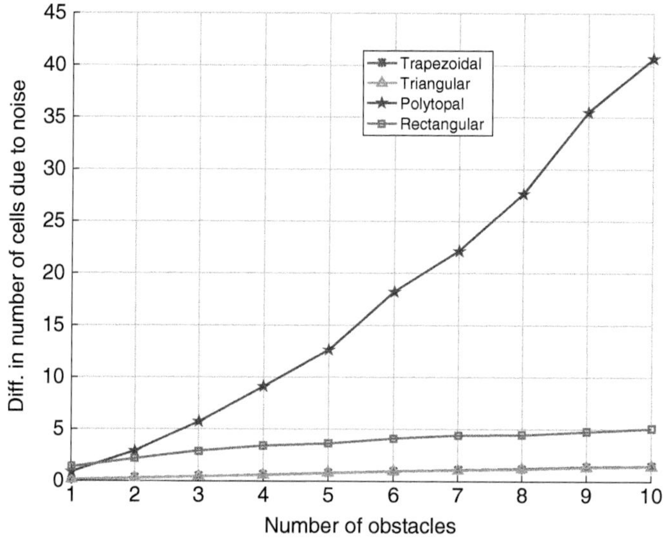

(h) Robustness measure as difference in number of cells due to noise affecting obstacle vertices.

Figure 3.9 *(Continued)*

greatest variability. For trapezoidal cells, this fact could be inferred from item (iii) of Subsection 3.4.1.

vi) Cell compactness (Figure 3.9(f)) is a measure of the shape quality of a cell [29]. The compactness of a cell is defined as the ratio $4\ \pi\ cell_{area}/cell^2_{perimeter}$. The compactness of a circle is 1 and the compactness of a very narrow polygon is close to 0. In planning motion of mobile robots by cell decompositions, it is desirable to avoid long and narrow cells; therefore a cell compactness close to 1 is beneficial. For the rectangular decomposition, all cells have the same compactness (because they are rectangles with the same ratio between sides). This fact is also visible in Figure 3.9(f) and in the standard deviation in Figure 3.9(g). In terms of compactness, the rectangular decomposition is desirable, followed by the polytopal one.

vii) Figure 3.9(h) presents the difference in the number of cells that result when the decomposition is created for the same environment, but after altering the obstacle vertices with a small noise. This cost criterion is a measure of the robustness of a partition with respect to small changes in the environment, e.g. noise due to a surveillance video camera. The polytopal decomposition proves to be the least robust one, as qualitatively inferred in item (iv) from Subsection 3.4.1.

viii) The percentage of small cells from a decomposition is illustrated in Figures 3.10(a) and 3.10(b). A small cell is considered a cell with an area smaller than an imposed threshold. This threshold was chosen as a percentage of the average obstacle area (in Figure 3.10(a)) and as a percentage of the environment area (in Figure 3.10(b)). Both cases show that the triangular and trapezoidal decompositions generally have a small percentage of small cells and the polytopal decomposition has a small percentage of small cells only for one or two obstacles. Again, the percentage of small cells for the rectangular decomposition should be mildly considered, as mentioned previously.

The presented tendencies for all cost criteria introduced in this section, together with the properties from Section 3.4.1, can serve as a way to decide what cell decomposition should be firstly chosen in searching for a solution. However, we recall that these comparisons rely solely on the cell decompositions, not on their usage for specific problems, and further details from the point of view of robot planning will be discussed in Chapter 5. Moreover, the generally small computation times for performing a decomposition and the ability to rerun a problem with a different decomposition (as, for example, in RMTool) give promising hypothesis in using the mentioned decompositions.

Table 3.2 summarizes some variations of the discussed criteria versus the number of obstacles for each decomposition type.

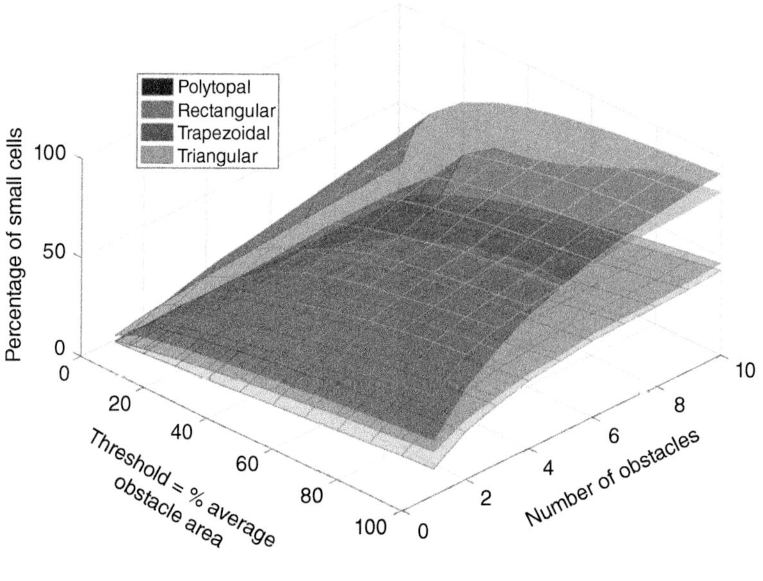

(a) Average obstacle size.

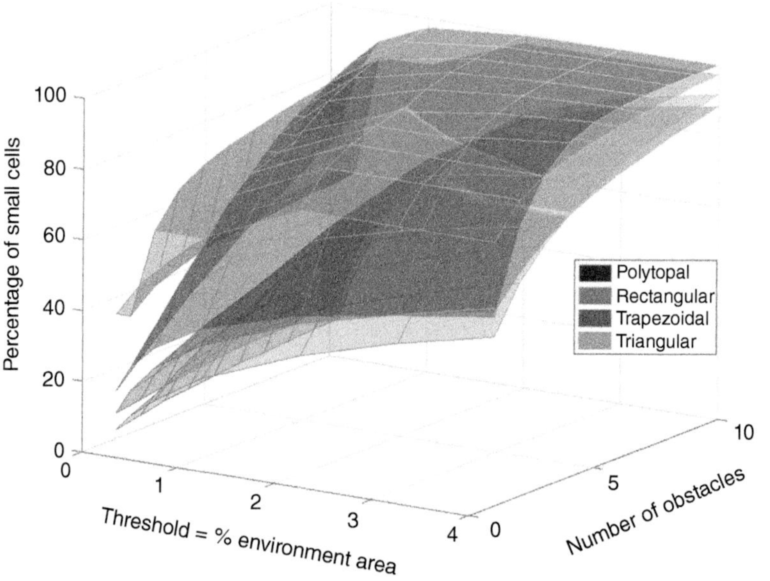

(b) Environment area.

Figure 3.10 Comparison of cell decomposition approaches. Percentage of small cells, obtained by considering thresholds.

Table 3.2 Trends of metrics with respect to the number of obstacles: cost criteria on rows and cell decomposition types on columns.

	Trapezoidal	Triangular	Polytopal	Rectangular
Number of cells	Small, slowly increases	Small, slowly increases	Quickly increases to large values	Average, slowly increases
Computation time	Small	Small	Quickly increases	Small
Number of adjacency relations	Small, slowly increases	Small, slowly increases	Quickly increases to large values	Average, slowly increases
Average cell area	Quickly decreases; large standard deviation	Quickly decreases; average standard deviation	Smallest for more than 3 obstacles	Average, with small variations
Average cell compactness	Smallest for more than 5 obstacles	Smallest for less than 5 obstacles	Average, with negligible variations	Largest, constant

3.5 Conclusions

This chapter presents different methods for constructing cell decompositions for an environment cluttered with polygonal obstacles or regions of interest. The planar environment is rectangular and the user-defined shapes are static and known. Four cell decompositions were considered, namely trapezoidal, triangular, polytopal, and rectangular. Each decomposition yields polygonal and convex cells, with the shape suggested by the decomposition name.

Section 3.2 discusses the hypothesis we consider for the environment and includes an overview of each decomposition, together with algorithmic pseudocodes and simple examples.

Section 3.3 presents the developed software implementations for these decompositions. All the functions were developed under Matlab and are included in RMTool; a user can also employ them outside the graphical interface. The triangular and polytopal decompositions are also extended to decomposing the defined regions (obstacles), rather than only the free space. The section includes various examples for manipulating the available functions, starting with the environment definitions and continuing with its decomposition and graphical representation.

Section 3.4 focuses on qualitative and quantitative comparisons among the four types of decompositions. The qualitative differentiations are mainly based on the underlying decomposition algorithms, while the quantitative comparison follows the trends of some metrics based on numerical experiments that use the software implementations. Tables 3.1 and 3.2 sketch the resemblances and differences in several aspects.

The remainder of the book will use the discussed cell decompositions for transposing a given mobile robot problem from the continuous world into a discrete event framework.

4

Discrete Event System Models

4.1 Introduction

Assuming a team of identical omnidirectional point robots evolving in a continuous space in which some obstacles or regions of interest may exist, the robots should cooperate in order to sequentially reach some destinations (possible with necessary synchronizations) for fulfilling a common task. In order to obtain paths for robots, the most common procedure is the use of mathematical (formal) discrete models. Since the space in which the robots are moving is continuous, a possible representation type is given by continuous models. These models can be easily obtained by writing the differential equations corresponding to the robot dynamics. However, these equations are in general non-linear, resulting in non-linear models. Furthermore, if one wants to model synchronizations in the presence of obstacles, the models become hybrid. Even if the hybrid-system community is quite large and many analysis and synthesis methods exist, the problem of computing paths on continuous space is difficult. In robotics, there exist results for the simple navigation problem for a robot (reaching a final destination by avoiding obstacles), for example using *navigation functions* [134], but there is no general result for high-level specifications and/or multi-robot systems. Notice that in this case, additional problems may appear, such as, for example, collision avoidance [71].

In order to tackle the complex problem explained above, one solution is to use approximations to discrete models. The difference between these models with respect to the continuous models is that the state-space is now discrete. Therefore, the state variables can take only discrete values. In the case of our application, the environment will be split into a finite number of cells (based on a cell decomposition algorithm) and to each cell a discrete state will be assigned. Independently by the exact position of the robot within the cell, it will be approximated to the same discrete state. Obviously, if the number of cells in which the environment is split is higher then the approximation will be better,

Path Planning of Cooperative Mobile Robots Using Discrete Event Models, First Edition.
Cristian Mahulea, Marius Kloetzer, and Ramón González.

but the computational complexity of computing paths and controllers for the robots will be higher. Thus, the size of the discrete model is directly linked to the accuracy of the discretization of the environment and, hence, to complexity and path accuracy.

The discrete models presented in this chapter are obtained based on the partition resulting from a cell decomposition algorithm described in the previous chapter. Section 4.2 presents a graph-based model that is obtained considering only the partition of the environment. By assuming some additional information of the robots as, for example, the robot dynamics, Section 4.3 introduces a transition system model of a robot and of a team of robots. In the case of a team of robots, the main drawback of this method is that when the number of robots is increased, the complexity of the model (the number of discrete nodes necessary to describe the system) also increases exponentially. In order to overcome this problem, Section 4.4 introduces Petri net models where the structure of the model will remain unchanged independently of the number of robots in the team. Section 4.5 presents a Petri net model that includes capacity constraints in some given cells that can be used to avoid collision in multi-robot systems. High-level specifications and preliminaries on Linear Temporal Logic are discussed in Sections 4.6 and 4.7, respectively.

4.2 Environment abstraction

Let us assume a 2-D rectangular environment that may contain convex obstacle regions and convex regions of interest. In this environment a team of identical robots will be deployed that should fulfill a given task. In order to compute paths for the robots, the first step is to obtain a model coding all possible movements that can be done by the robots. Discrete models will be used and are defined based on a graph abstraction of the environment.

Before formally defining the graph abstraction, let us consider an example. In particular, let us assume the rectangular environment in Figure 4.1 that contains two obstacles (black regions in the figure). By applying the polytopal cell decomposition algorithm presented in Section 3.2.4, it has been divided into six convex cells $c_1, c_2, \dots, c_6$. In order to define the graph abstraction of this partitioned environment, let us define for each cell obtained after decomposition a discrete node in the graph. In order to simplify the notation, let us denote the discrete nodes with the same symbols as the name of cells. Hence, the environment in Figure 4.1 will be abstracted by a graph with six discrete nodes, $C = \{c_1, c_2, c_3, c_4, c_5, c_6\}$.

The corresponding edges of the graph are obtained based on the adjacency relation. If two cells have a common border, then in the graph abstraction the corresponding nodes will be connected by an undirected arc. For example, there exists

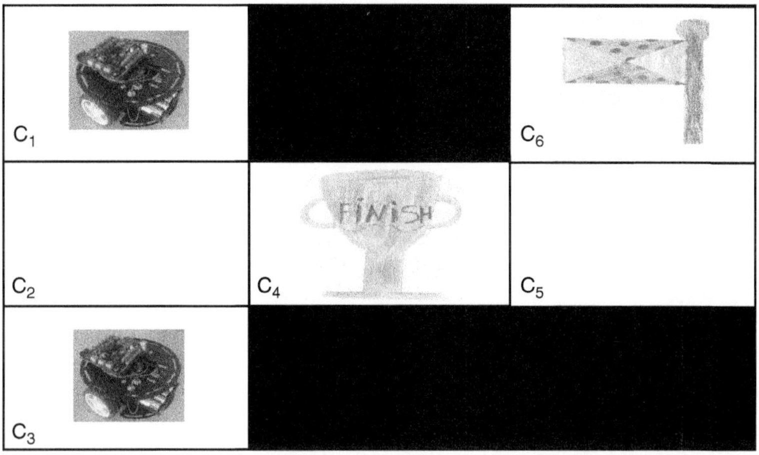

Figure 4.1 An example of a discretized environment. A discretized environment with two obstacles and two regions of interest in which two robots are deployed.

Figure 4.2 Graph abstraction of an environment. Graph abstraction of the discretized environment depicted in Figure 4.1.

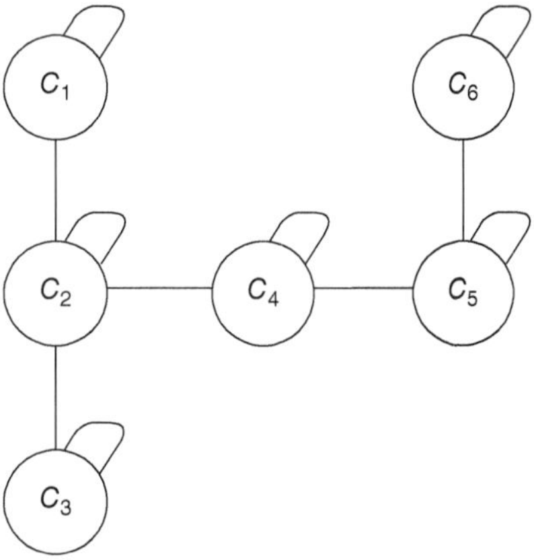

an arc between c_1 and c_2 because they are adjacent cells but there exists no arc between c_1 and c_4 because they are not adjacent (they do not share a whole edge). The corresponding graph abstraction of the discretized environment of Figure 4.1 is given in Figure 4.2.

Some cells from the environment should be distinguished from the others because they will be used to define the task and to obtain paths. For example, assume that the two robots should arrive in cell c_4 and one robot should cross cell c_6 before stopping in c_4. In order to be able to do this, we will define a set of regions of interest. Since one robot should cross c_4, one region of interest will be $O_1 = \{c_4\}$ while the other region of interest will be $O_2 = \{c_6\}$ corresponding to the final cell. The set of all regions of interest will be called O, which for this example will be $O = \{O_1, O_2\}$. Notice that these regions of interest are defined in general before the partition, and during the cell decomposition they could be divided into several cells; see, for example, Figure 3.7.

Definition 4.1 A graph abstraction of the environment is a tuple $G = \langle C, E, O \rangle$, where:

- C is a finite set of discrete nodes representing the cells in the environment where robots can travel;
- $E \subseteq C \times C$ is the set of undirected edges, where $(c_i, c_j) \in E$ means that the regions modeled by nodes c_i and c_j are adjacent (have a common border); furthermore, we assume that $(c_i, c_i) \in E$ for all $c_i \in C$ (as in Figure 4.2);
- $O = \{O_1, O_2, \ldots, O_{|O|}\}$ is the finite set of the regions of interest. In particular, any $O_i \in O$ is the set of nodes corresponding to the region of interest i, i.e. $O_i \subseteq C$. In some cases, O_i is a singleton containing only one node, but a convex region of interest may be divided into cells during the cell decomposition as discussed in Chapter 3. Moreover, two regions of interest may or may not be disjoint.

Given a graph G, the adjacency relation defined by the set E can be represented also by a matrix $\boldsymbol{A} \in \mathbb{R}^{|C| \times |C|}$ defined as follows:

$$\boldsymbol{A}[i,j] = \begin{cases} 1, & \text{if } (c_i, c_j) \in E \\ 0, & \text{otherwise.} \end{cases} \tag{4.1}$$

Notice that since the graph is undirected, adjacency matrix $\boldsymbol{A}$ is symmetric. Additionally, the diagonal elements of $\boldsymbol{A}$ are all equal to 1.

Considering again the environment given in Figure 4.2, the graph abstraction of the discretized environment is given by the following sets:

- $C = \{c_1, c_2, c_3, c_4, c_5, c_6\}$;
- $E = \{(c_1, c_1), (c_1, c_2), (c_2, c_2), (c_2, c_3), (c_2, c_4), (c_3, c_3), (c_4, c_4), (c_4, c_5), (c_5, c_5), (c_5, c_6), (c_6, c_6)\}$;
- $O = \{O_1, O_2\}$, where $O_1 = \{c_4\}$ and $O_2 = \{c_6\}$.

Notice that in this case $O_1 \cap O_2 = \emptyset$. The adjacency matrix is the following:

$$\boldsymbol{A} = \begin{bmatrix} 1 & 1 & 0 & 0 & 0 & 0 \\ 1 & 1 & 1 & 1 & 0 & 0 \\ 0 & 1 & 1 & 0 & 0 & 0 \\ 0 & 1 & 0 & 1 & 1 & 0 \\ 0 & 0 & 0 & 1 & 1 & 1 \\ 0 & 0 & 0 & 0 & 1 & 1 \end{bmatrix} \tag{4.2}$$

In RMTool, after defining an environment, it is possible to export to the workspace the set of discrete states and the adjacency matrix of the partitioned environment by using the option *Export to Workspace → Environment* from the menu. Automatically, an input dialog box is opened in order to select the name of the following variables: (i) the set of discrete states, (ii) the set of regions of interest, and (iii) the adjacency matrix. In this way, the user can employ the graph corresponding to the displayed environment in order to apply other algorithms and/or study other properties.

4.3 Transition system models

Given an environment and a set of regions of interest O, to each region of interest $O_i \in O$ we assign an output y_i. The robot motion tasks will be defined on the set $Y = \{y_1, y_2, \dots, y_{|O|}\}$, each symbol from Y referring to a region of interest that should be visited or avoided by the robot(s).

4.3.1 Single robot case

Let us assume only one robot in the environment and let us define a transition system that will model its movement capabilities. The transition system is constructed based on:

- (i) the graph description of the environment given in Definition 4.1.
- (ii) the robot dynamics implying a refinement of the graph description from (i) by removing transitions that are not possible due to the non-existence of control laws to enforce the transition. In order to check the existence of the control laws for arbitrary robot dynamics, control design techniques for particular partitions can be used [12, 14, 91].
- (iii) a satisfaction map that shows the regions of interest that are active when the robot is inside a particular cell (under the note that each region of interest is equal to a union of partition cells).

Since in this book we assume omnidirectional (fully actuated) point robots (see Section 1.5), condition (ii) does not present any problem and the transition system will be constructed mainly based on the graph abstraction of the environment (item (i)) and each region represented as a union of cells (item (iii)).

Let us consider again the environment in Figure 4.1. The transition system modeling the movement capabilities of the robot initially placed in c_1 has 6 states as the graph abstraction of the environment (see Figure 4.2). Since initially the robot is placed in c_1, this will be the initial state of the transition system. Let us assume that some simple control laws are the following: *up, down, left, right*, and *stop*. For

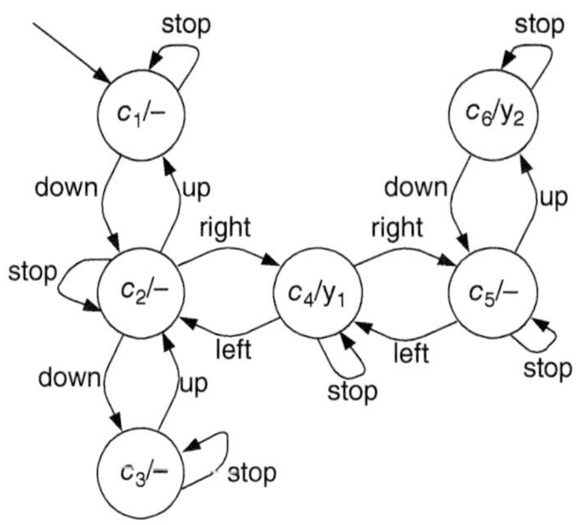

Figure 4.3 Example of a transition system model of a robot. Transition system model of the robot placed in cell c_1 in the discretized environment of Figure 4.1. The arcs are labeled by the inputs (control laws), the initial state is c_1 while the states labeled with '-' means empty output, i.e. $\emptyset$.

example, if *up* is applied, the robot moves to the nearest cell placed to the north of the current one. From any point in c_1, the robot can go to c_2 if input *down* is applied. Hence, there exists a direct arc (a transition) from c_1 to c_2 labeled as *down*. If the robot moves from c_1 to c_2, the state of the system changes from c_1 to c_2.

The task of the robot will be defined in relation to some regions of the environment, called regions of interest. For example, assume that the robot should go to $O_2 = \{c_6\}$ and then go and stop in $O_1 = \{c_4\}$. In order to model this, we assign (as mentioned) outputs to the regions of interest, namely output y_i will correspond to region $O_i, i = 1, \ldots, |O|$. If the robot is inside region c_4 (or equivalently that the current state of the transition system is c_4), then the output y_1 is active, if it is inside c_6 the output y_2 is active, and there is no output if the robot is inside any other region except these two. The output function is denoted by h and gives for each cell the outputs that are active if the robot is inside that cell. In our example, $h(c_4) = y_1$, $h(c_6) = y_2$, and $h(c_1) = h(c_2) = h(c_3) = h(c_5) = \emptyset$, where $\emptyset$ means empty output, i.e. the robot is not in a region of interest. The complete transition system is sketched in Figure 4.3.

A path for the robot is a sequence of cells that will be crossed by the robot and corresponds to a path or run in the transition system. Assuming the task described before, a satisfying path is $\rho = c_1c_2c_4c_5c_6c_5c_4$, which can be followed from c_1 by the sequence of inputs: *down, right, right, up, down, left, stop*. This path generates an output word denoted $h(\rho) = h(c_1)h(c_2)h(c_4)h(c_5)h(c_6)h(c_5)h(c_4) = y_1y_2y_1$. According to the output word, the task is fulfilled if the robot follows this path.

As mentioned, we consider omnidirectional point robots; hence there always exists a control law to steer the robot from one cell to an adjacent cell. For this reason, and in order to simplify the notation, we will not include explicitly in the definition of the transition systems the inputs that make possible the robot movements. Once the path is obtained, the sequence of inputs (or the sequence of control laws) that should be applied to the robot can be easily obtained. A formal definition of the transition system is given next.

Definition 4.2 Given an environment represented by an undirected graph $G = \langle \overline{C}, \overline{E}, O \rangle$ (where $\overline{C}$ is the set of discrete nodes, $\overline{E}$ is the set of edges, and O the set of regions of interests), with $\boldsymbol{A}$ the adjacency matrix, and an omnidirectional point robot that is evolving inside the environment, the *transition system* of the robot is a tuple $T_S = \langle C, c_0, E, H, h \rangle$, where:

- C is the finite number of states, one state for each cell of the environment, i.e. $C = \overline{C}$;
- c_0 is the initial state (corresponding to the cell containing the initial deployment of the robot);
- $E \subseteq C \times C$ is the transition relation, where $(c_i, c_j) \in E$ if $\boldsymbol{A}[i,j] = 1$, meaning that the robot can be steered from any initial condition in c_i to the adjacent cell c_j without visiting any other neighboring cells in finite time, or it can be kept inside cell c_i for an indefinite time if $c_i = c_j$;
- $h : C \to 2^Y$ is the satisfaction function (with $Y = \{y_1, y_2, \dots, y_{|O|}\}$), which returns outputs (observations of states of T_S) yielded by the robot. Recall that output y_i corresponds to the region of interest $O_i \in O$, and more symbols from Y (e.g. an element of 2^Y) are needed if two or more regions partially overlap and the robot is located in their intersection. The $\emptyset$ element of the power set corresponds to empty output (no region of interest observed/visited) [1]. With a slight abuse of notation, y_i is also regarded as a Boolean variable and it can be said that the output y_i is active (or that the observation y_i is satisfied), i.e. $y_i = 1$, if the robot is inside a cell from O_i; otherwise $y_i = 0$. Since the regions of interest may overlap, a possible output can be seen as a conjunction of elements y_i; e.g. if $c_i \in O_j \cap O_k$ then if the robot is inside c_i, observation $\{y_j, y_k\}$ (conjunction $y_j \wedge y_k$) is active. Therefore, $h(c_i)$ is the set of all outputs active in cell c_i, with $h(c_i) = \emptyset$ if cell c_i is not belonging to any region of interest.
- $H = \bigcup_{c \in C} h(c)$ is the set of all possible observations of T_S, being $H \subseteq 2^Y$.

Transition systems that we use are deterministic, because any feasible transition in the current state can be enforced by imposing a specific robot control law. A *run* (or path) of T_S is an infinite sequence $\rho = c_0 c_1 c_2 \dots$, with the property that $(c_i, c_{i+1}) \in E$, $\forall i \geq 0$. Furthermore, a run induces (through function h) an output *word*, which is the observed sequence of elements from H. The set of possible observations H will be the same with the set of inputs of a Büchi automaton B (defined later in Section 4.7), thus making that an output word of T_S can be an

1 In some studies, notation ε is also used for an empty observation.

input word for B. Therefore, one can make the connection between robot paths and satisfaction of LTL specifications, as will be described in Chapter 5.

Given a graph abstraction $G = \langle \overline{C}, \overline{E}, O \rangle$ and an omnidirectional point robot, the transition system model T_S is obtained by Algorithm 4.1, in which, for a set S, notation $|S|$ denotes the cardinality of the set, i.e. the number of elements of S. The set of discrete states of T_S is initialized with the set of nodes of the graph (step 1); an output symbol y_i is assigned to each region of interest O_i (step 2); and the set of edges E is initialized to the empty set (step 3). The initial state of T_S is equal to the discrete state corresponding to the cell in which the robot is initially deployed (step 4). The set of edges is created based on the adjacency matrix $\boldsymbol{A}$ by adding a new edge (c_i, c_j) for each element of $\boldsymbol{A}[i,j]$ equal to 1 (step 9). Furthermore, the satisfaction function γ is initialized to e empty set (step 6) and updated according to sets in O (steps 13 and 15), finally yielding the set of possible outputs H (step 16).

Going back to the environment in Figure 4.1, the transition system corresponding to the robot initially placed in c_1 is $T_S = \langle C, c_0, E, H, h \rangle$, where

Algorithm 4.1: Construct the transition system T_S

Input: Environment model $G = \langle \overline{C}, \overline{E}, O \rangle$, $\boldsymbol{A}$, initial position of the robot
Output: Robot model $T_S = \langle C, c_0, E, H, h \rangle$

1 Let $C = \overline{C}$;
2 Let $E = \emptyset$;
3 Let y_i be the output symbol corresponding to O_i, $i = 1, \ldots, |O|$;
4 Let c_0 be the region from C containing the initial position of the robot;
5 Let $h(c_i) = \emptyset, \forall c_i \in C$;
6 **for** $i = 1$ *to* $|C| - 1$ **do**
7   **for** $j = i + 1$ *to* $|C|$ **do**
8   **if** $(\boldsymbol{A}[i,j] == 1)$ **then**
9   Let $E = E \cup \{(c_i, c_j), (c_j, c_i)\}$;

10 **forall** $O_i \in O$ **do**
11   **forall** $c_j \in O_i$ **do**
12   **if** $(h(c_j) == \emptyset)$ **then**
13             $h(c_j) = \{y_i\}$;
14   **else**
15   Let $h(c_j) = h(c_j) \cup \{y_i\}$;

16 Let $H = \bigcup_{c_i \in C} h(c_i)$;

- $C = \{c_1, c_2, c_3, c_4, c_5, c_6\}$;
- $c_0 = c_1$;
- $E = \{(c_1, c_1),\ (c_1, c_2),\ (c_2, c_1),\ (c_2, c_2),\ (c_2, c_3),\ (c_2, c_4),\ (c_3, c_2),\ (c_3, c_3),\ (c_4, c_2), (c_4, c_4), (c_4, c_5), (c_5, c_4), (c_5, c_5), (c_5, c_6), (c_6, c_5), (c_6, c_6)\}$;
- $H = \{y_1, y_2, \emptyset\}$;
- $h(c_1) = h(c_2) = h(c_3) = h(c_5) = \emptyset$, $h(c_4) = y_1$ and $h(c_6) = y_2$.

4.3.2 Multi-robot case

Assume now a team $R = \{r_1, r_2, \ldots, r_{|R|}\}$ of identical robots evolving in the same environment. Since the robots are identical, each robot i is modeled by the same transition system $T_i = \langle C, c_{0i}, E, H, h\rangle$. The only difference is given by the initial state that corresponds to the initial cell c_{0i} of robot i.

Informally, we construct a team transition system T_T that captures the possible behaviors of the whole team of $|R|$ robots. Thus, states of T_T are $|R|$-tuples in which the ith element shows the location of robot i. Transitions between states correspond to one or more robots changing their current cell from C. When a transition implies that more robots change their cells, these changes are assumed to be synchronous, meaning that the moving robots cross the borders between cells at the same time. Such a behavior can be enforced by waiting modes enabled for robots that arrive faster at the border of the next cell they should visit.

Considering again the environment in Figure 4.1, the transition system of the robot initially placed at c_1 has been explained before and was shown in Figure 4.3. The transition system of the robot initially placed at c_3 is similar to the previous one, the only difference being in the initial state that is equal now to c_3 instead of c_1. In order to consider the transition system of the team, we notice that now the state of the model will be a pair $\langle c_i, c_j\rangle$, where c_i is the actual state of first robot and c_j the actual state of the second robot. Assuming for the second robot the same inputs as before (i.e. *up, down, left, right*, and *stop*), the transition system of the team composed by both robots is given as follows:

- The set of states is $C_T = \{\langle c_1, c_1\rangle, \langle c_1, c_2\rangle, \langle c_1, c_3\rangle, \langle c_1, c_4\rangle, \langle c_1, c_5\rangle, \langle c_1, c_6\rangle, \langle c_2, c_1\rangle, \langle c_2, c_2\rangle, \langle c_2, c_3\rangle, \langle c_2, c_4\rangle, \langle c_2, c_5\rangle, \langle c_2, c_6\rangle, \langle c_3, c_1\rangle, \langle c_3, c_2\rangle, \langle c_3, c_3\rangle, \langle c_3, c_4\rangle, \langle c_3, c_5\rangle, \langle c_3, c_6\rangle, \langle c_4, c_1\rangle, \langle c_4, c_2\rangle, \langle c_4, c_3\rangle, \langle c_4, c_4\rangle, \langle c_4, c_5\rangle, \langle c_4, c_6\rangle, \langle c_5, c_1\rangle, \langle c_5, c_2\rangle, \langle c_5, c_3\rangle, \langle c_5, c_4\rangle, \langle c_5, c_5\rangle, \langle c_5, c_6\rangle, \langle c_6, c_1\rangle, \langle c_6, c_2\rangle, \langle c_6, c_3\rangle, \langle c_6, c_4\rangle, \langle c_6, c_5\rangle, \langle c_6, c_6\rangle\}$, where in each state the first element represents the cell of the first robot and the second element is the cell of the second robot. The number of cells is equal to 6; hence the total number of states for the team is $6^2 = 36$.
- The initial state is $\langle c_1, c_3\rangle$.
- The set of edges E_T is given by the following list, where each of the 36 dashes marks the outgoing transitions from one state:

- $(\langle c_1,c_1\rangle,\langle c_1,c_1\rangle)$, $(\langle c_1,c_1\rangle,\langle c_2,c_1\rangle)$, $(\langle c_1,c_1\rangle,\langle c_1,c_2\rangle)$, $(\langle c_1,c_1\rangle,\langle c_2,c_2\rangle)$. These are output edges of state $\langle c_1,c_1\rangle$. The first edge $(\langle c_1,c_1\rangle,\langle c_1,c_1\rangle)$ corresponds to a self-loop in state $\langle c_1,c_1\rangle$ and can be executed with input *stop* for both robots, while edge $(\langle c_1,c_1\rangle,\langle c_2,c_2\rangle)$ corresponds to the synchronous movement of both robots from cell c_1 to cell c_2 and can be executed (when both robots are in c_1) by applying input *down* to both robots and ensuring synchronous entering in the next cell.
- $(\langle c_1,c_2\rangle,\langle c_1,c_2\rangle)$, $(\langle c_1,c_2\rangle,\langle c_2,c_2\rangle)$, $(\langle c_1,c_2\rangle,\langle c_1,c_1\rangle)$, $(\langle c_1,c_2\rangle,\langle c_1,c_3\rangle)$, $(\langle c_1,c_2\rangle,\langle c_1,c_4\rangle)$, $(\langle c_1,c_2\rangle,\langle c_2,c_1\rangle)$, $(\langle c_1,c_2\rangle,\langle c_2,c_3\rangle)$, $(\langle c_1,c_2\rangle,\langle c_2,c_4\rangle)$;
- $(\langle c_1,c_3\rangle,\langle c_1,c_3\rangle)$, $(\langle c_1,c_3\rangle,\langle c_2,c_3\rangle)$, $(\langle c_1,c_3\rangle,\langle c_1,c_2\rangle)$, $(\langle c_1,c_3\rangle,\langle c_2,c_2\rangle)$;
- $(\langle c_1,c_4\rangle,\langle c_1,c_4\rangle)$, $(\langle c_1,c_4\rangle,\langle c_2,c_4\rangle)$, $(\langle c_1,c_4\rangle,\langle c_1,c_2\rangle)$, $(\langle c_1,c_4\rangle,\langle c_1,c_5\rangle)$, $(\langle c_1,c_4\rangle,\langle c_2,c_2\rangle)$, $(\langle c_1,c_4\rangle,\langle c_2,c_5\rangle)$;
- $(\langle c_1,c_5\rangle,\langle c_1,c_5\rangle)$, $(\langle c_1,c_5\rangle,\langle c_2,c_5\rangle)$, $(\langle c_1,c_5\rangle,\langle c_1,c_4\rangle)$, $(\langle c_1,c_5\rangle,\langle c_1,c_6\rangle)$, $(\langle c_1,c_5\rangle,\langle c_2,c_4\rangle)$, $(\langle c_1,c_5\rangle,\langle c_2,c_6\rangle)$;
- $(\langle c_1,c_6\rangle,\langle c_1,c_6\rangle)$, $(\langle c_1,c_6\rangle,\langle c_2,c_6\rangle)$, $(\langle c_1,c_6\rangle,\langle c_1,c_5\rangle)$, $(\langle c_1,c_6\rangle,\langle c_2,c_5\rangle)$;
- $(\langle c_2,c_1\rangle,\langle c_2,c_1\rangle)$, $(\langle c_2,c_1\rangle,\langle c_1,c_1\rangle)$, $(\langle c_2,c_1\rangle,\langle c_3,c_1\rangle)$, $(\langle c_2,c_1\rangle,\langle c_4,c_1\rangle)$, $(\langle c_2,c_1\rangle,\langle c_2,c_2\rangle)$, $(\langle c_2,c_1\rangle,\langle c_1,c_2\rangle)$, $(\langle c_2,c_1\rangle,\langle c_3,c_2\rangle)$, $(\langle c_2,c_1\rangle,\langle c_4,c_2\rangle)$;
- $(\langle c_2,c_2\rangle,\langle c_2,c_2\rangle)$, $(\langle c_2,c_2\rangle,\langle c_1,c_2\rangle)$, $(\langle c_2,c_2\rangle,\langle c_3,c_2\rangle)$, $(\langle c_2,c_2\rangle,\langle c_4,c_2\rangle)$, $(\langle c_2,c_2\rangle,\langle c_2,c_1\rangle)$, $(\langle c_2,c_2\rangle,\langle c_2,c_3\rangle)$, $(\langle c_2,c_2\rangle,\langle c_2,c_4\rangle)$, $(\langle c_2,c_2\rangle,\langle c_1,c_1\rangle)$, $(\langle c_2,c_2\rangle,\langle c_1,c_3\rangle)$, $(\langle c_2,c_2\rangle,\langle c_1,c_4\rangle)$, $(\langle c_2,c_2\rangle,\langle c_3,c_1\rangle)$, $(\langle c_2,c_2\rangle,\langle c_3,c_3\rangle)$, $(\langle c_2,c_2\rangle,\langle c_3,c_4\rangle)$, $(\langle c_2,c_2\rangle,\langle c_4,c_1\rangle)$, $(\langle c_2,c_2\rangle,\langle c_4,c_3\rangle)$, $(\langle c_2,c_2\rangle,\langle c_4,c_4\rangle)$;
- $(\langle c_2,c_3\rangle,\langle c_2,c_3\rangle)$, $(\langle c_2,c_3\rangle,\langle c_1,c_3\rangle)$, $(\langle c_2,c_3\rangle,\langle c_3,c_3\rangle)$, $(\langle c_2,c_3\rangle,\langle c_4,c_3\rangle)$, $(\langle c_2,c_3\rangle,\langle c_2,c_2\rangle)$, $(\langle c_2,c_3\rangle,\langle c_1,c_2\rangle)$, $(\langle c_2,c_3\rangle,\langle c_3,c_2\rangle)$, $(\langle c_2,c_3\rangle,\langle c_4,c_2\rangle)$;
- $(\langle c_2,c_4\rangle,\langle c_2,c_4\rangle)$, $(\langle c_2,c_4\rangle,\langle c_1,c_4\rangle)$, $(\langle c_2,c_4\rangle,\langle c_3,c_4\rangle)$, $(\langle c_2,c_4\rangle,\langle c_4,c_4\rangle)$, $(\langle c_2,c_4\rangle,\langle c_2,c_2\rangle)$, $(\langle c_2,c_4\rangle,\langle c_2,c_5\rangle)$, $(\langle c_2,c_4\rangle,\langle c_1,c_2\rangle)$, $(\langle c_2,c_4\rangle,\langle c_1,c_5\rangle)$, $(\langle c_2,c_4\rangle,\langle c_3,c_2\rangle)$, $(\langle c_2,c_4\rangle,\langle c_3,c_5\rangle)$, $(\langle c_2,c_4\rangle,\langle c_4,c_2\rangle)$, $(\langle c_2,c_4\rangle,\langle c_4,c_5\rangle)$;
- $(\langle c_2,c_5\rangle,\langle c_2,c_5\rangle)$, $(\langle c_2,c_5\rangle,\langle c_1,c_5\rangle)$, $(\langle c_2,c_5\rangle,\langle c_3,c_5\rangle)$, $(\langle c_2,c_5\rangle,\langle c_4,c_5\rangle)$, $(\langle c_2,c_5\rangle,\langle c_2,c_4\rangle)$, $(\langle c_2,c_5\rangle,\langle c_2,c_6\rangle)$, $(\langle c_2,c_5\rangle,\langle c_1,c_4\rangle)$, $(\langle c_2,c_5\rangle,\langle c_1,c_6\rangle)$, $(\langle c_2,c_5\rangle,\langle c_3,c_4\rangle)$, $(\langle c_2,c_5\rangle,\langle c_3,c_6\rangle)$, $(\langle c_2,c_5\rangle,\langle c_4,c_4\rangle)$, $(\langle c_2,c_5\rangle,\langle c_4,c_6\rangle)$;
- $(\langle c_2,c_6\rangle,\langle c_2,c_6\rangle)$, $(\langle c_2,c_6\rangle,\langle c_1,c_6\rangle)$, $(\langle c_2,c_6\rangle,\langle c_3,c_6\rangle)$, $(\langle c_2,c_6\rangle,\langle c_4,c_6\rangle)$, $(\langle c_2,c_6\rangle,\langle c_2,c_5\rangle)$, $(\langle c_2,c_6\rangle,\langle c_1,c_5\rangle)$, $(\langle c_2,c_6\rangle,\langle c_3,c_5\rangle)$, $(\langle c_2,c_6\rangle,\langle c_4,c_5\rangle)$;
- $(\langle c_3,c_1\rangle,\langle c_3,c_1\rangle)$, $(\langle c_3,c_1\rangle,\langle c_2,c_1\rangle)$, $(\langle c_3,c_1\rangle,\langle c_3,c_2\rangle)$, $(\langle c_3,c_1\rangle,\langle c_2,c_2\rangle)$;
- $(\langle c_3,c_2\rangle,\langle c_3,c_2\rangle)$, $(\langle c_3,c_2\rangle,\langle c_2,c_2\rangle)$, $(\langle c_3,c_2\rangle,\langle c_3,c_1\rangle)$, $(\langle c_3,c_2\rangle,\langle c_3,c_3\rangle)$, $(\langle c_3,c_2\rangle,\langle c_3,c_4\rangle)$, $(\langle c_3,c_2\rangle,\langle c_2,c_1\rangle)$, $(\langle c_3,c_2\rangle,\langle c_2,c_3\rangle)$, $(\langle c_3,c_2\rangle,\langle c_2,c_4\rangle)$;
- $(\langle c_3,c_3\rangle,\langle c_3,c_3\rangle)$, $(\langle c_3,c_3\rangle,\langle c_2,c_3\rangle)$, $(\langle c_3,c_3\rangle,\langle c_3,c_2\rangle)$, $(\langle c_3,c_3\rangle,\langle c_2,c_2\rangle)$;
- $(\langle c_3,c_4\rangle,\langle c_3,c_4\rangle)$, $(\langle c_3,c_4\rangle,\langle c_2,c_4\rangle)$, $(\langle c_3,c_4\rangle,\langle c_3,c_2\rangle)$, $(\langle c_3,c_4\rangle,\langle c_3,c_5\rangle)$, $(\langle c_3,c_4\rangle,\langle c_2,c_2\rangle)$, $(\langle c_3,c_4\rangle,\langle c_2,c_5\rangle)$;
- $(\langle c_3,c_5\rangle,\langle c_3,c_5\rangle)$, $(\langle c_3,c_5\rangle,\langle c_2,c_5\rangle)$, $(\langle c_3,c_5\rangle,\langle c_3,c_4\rangle)$, $(\langle c_3,c_5\rangle,\langle c_3,c_6\rangle)$, $(\langle c_3,c_5\rangle,\langle c_2,c_4\rangle)$, $(\langle c_3,c_5\rangle,\langle c_2,c_6\rangle)$;
- $(\langle c_3,c_6\rangle,\langle c_3,c_6\rangle)$, $(\langle c_3,c_6\rangle,\langle c_2,c_6\rangle)$, $(\langle c_3,c_6\rangle,\langle c_3,c_5\rangle)$, $(\langle c_3,c_6\rangle,\langle c_2,c_5\rangle)$;

- $(\langle c_4,c_1\rangle,\langle c_4,c_1\rangle)$, $(\langle c_4,c_1\rangle,\langle c_2,c_1\rangle)$, $(\langle c_4,c_1\rangle,\langle c_5,c_1\rangle)$, $(\langle c_4,c_1\rangle,\langle c_4,c_2\rangle)$, $(\langle c_4,c_1\rangle,\langle c_2,c_2\rangle)$, $(\langle c_4,c_1\rangle,\langle c_5,c_2\rangle)$;
- $(\langle c_4,c_2\rangle,\langle c_4,c_2\rangle)$, $(\langle c_4,c_2\rangle,\langle c_2,c_2\rangle)$, $(\langle c_4,c_2\rangle,\langle c_5,c_2\rangle)$, $(\langle c_4,c_2\rangle,\langle c_4,c_1\rangle)$, $(\langle c_4,c_2\rangle,\langle c_4,c_3\rangle)$, $(\langle c_4,c_2\rangle,\langle c_4,c_4\rangle)$, $(\langle c_4,c_2\rangle,\langle c_2,c_1\rangle)$, $(\langle c_4,c_2\rangle,\langle c_2,c_3\rangle)$, $(\langle c_4,c_2\rangle,\langle c_2,c_4\rangle)$, $(\langle c_4,c_2\rangle,\langle c_5,c_1\rangle)$, $(\langle c_4,c_2\rangle,\langle c_5,c_3\rangle)$, $(\langle c_4,c_2\rangle,\langle c_5,c_4\rangle)$;
- $(\langle c_4,c_3\rangle,\langle c_4,c_3\rangle)$, $(\langle c_4,c_3\rangle,\langle c_2,c_3\rangle)$, $(\langle c_4,c_3\rangle,\langle c_5,c_3\rangle)$, $(\langle c_4,c_3\rangle,\langle c_4,c_2\rangle)$, $(\langle c_4,c_3\rangle,\langle c_2,c_2\rangle)$, $(\langle c_4,c_3\rangle,\langle c_5,c_2\rangle)$;
- $(\langle c_4,c_4\rangle,\langle c_4,c_4\rangle)$, $(\langle c_4,c_4\rangle,\langle c_2,c_4\rangle)$, $(\langle c_4,c_4\rangle,\langle c_5,c_4\rangle)$, $(\langle c_4,c_4\rangle,\langle c_4,c_2\rangle)$, $(\langle c_4,c_4\rangle,\langle c_4,c_5\rangle)$, $(\langle c_4,c_4\rangle,\langle c_2,c_2\rangle)$, $(\langle c_4,c_4\rangle,\langle c_2,c_5\rangle)$, $(\langle c_4,c_4\rangle,\langle c_5,c_2\rangle)$, $(\langle c_4,c_4\rangle,\langle c_5,c_5\rangle)$;
- $(\langle c_4,c_5\rangle,\langle c_4,c_5\rangle)$, $(\langle c_4,c_5\rangle,\langle c_2,c_5\rangle)$, $(\langle c_4,c_5\rangle,\langle c_5,c_5\rangle)$, $(\langle c_4,c_5\rangle,\langle c_4,c_4\rangle)$, $(\langle c_4,c_5\rangle,\langle c_4,c_6\rangle)$, $(\langle c_4,c_5\rangle,\langle c_2,c_4\rangle)$, $(\langle c_4,c_5\rangle,\langle c_2,c_6\rangle)$, $(\langle c_4,c_5\rangle,\langle c_5,c_4\rangle)$, $(\langle c_4,c_5\rangle,\langle c_5,c_6\rangle)$;
- $(\langle c_4,c_6\rangle,\langle c_4,c_6\rangle)$, $(\langle c_4,c_6\rangle,\langle c_2,c_6\rangle)$, $(\langle c_4,c_6\rangle,\langle c_5,c_6\rangle)$, $(\langle c_4,c_6\rangle,\langle c_4,c_5\rangle)$, $(\langle c_4,c_6\rangle,\langle c_2,c_5\rangle)$, $(\langle c_4,c_6\rangle,\langle c_5,c_5\rangle)$;
- $(\langle c_5,c_1\rangle,\langle c_5,c_1\rangle)$, $(\langle c_5,c_1\rangle,\langle c_4,c_1\rangle)$, $(\langle c_5,c_1\rangle,\langle c_6,c_1\rangle)$, $(\langle c_5,c_1\rangle,\langle c_5,c_2\rangle)$, $(\langle c_5,c_1\rangle,\langle c_4,c_2\rangle)$, $(\langle c_5,c_1\rangle,\langle c_6,c_2\rangle)$;
- $(\langle c_5,c_2\rangle,\langle c_5,c_2\rangle)$, $(\langle c_5,c_2\rangle,\langle c_4,c_2\rangle)$, $(\langle c_5,c_2\rangle,\langle c_6,c_2\rangle)$, $(\langle c_5,c_2\rangle,\langle c_5,c_1\rangle)$, $(\langle c_5,c_2\rangle,\langle c_5,c_3\rangle)$, $(\langle c_5,c_2\rangle,\langle c_5,c_4\rangle)$, $(\langle c_5,c_2\rangle,\langle c_4,c_1\rangle)$, $(\langle c_5,c_2\rangle,\langle c_4,c_3\rangle)$, $(\langle c_5,c_2\rangle,\langle c_4,c_4\rangle)$, $(\langle c_5,c_2\rangle,\langle c_6,c_1\rangle)$, $(\langle c_5,c_2\rangle,\langle c_6,c_3\rangle)$, $(\langle c_5,c_2\rangle,\langle c_6,c_4\rangle)$;
- $(\langle c_5,c_3\rangle,\langle c_5,c_3\rangle)$, $(\langle c_5,c_3\rangle,\langle c_4,c_3\rangle)$, $(\langle c_5,c_3\rangle,\langle c_6,c_3\rangle)$, $(\langle c_5,c_3\rangle,\langle c_5,c_2\rangle)$, $(\langle c_5,c_3\rangle,\langle c_4,c_2\rangle)$, $(\langle c_5,c_3\rangle,\langle c_6,c_2\rangle)$;
- $(\langle c_5,c_4\rangle,\langle c_5,c_4\rangle)$, $(\langle c_5,c_4\rangle,\langle c_4,c_4\rangle)$, $(\langle c_5,c_4\rangle,\langle c_6,c_4\rangle)$, $(\langle c_5,c_4\rangle,\langle c_5,c_2\rangle)$, $(\langle c_5,c_4\rangle,\langle c_5,c_5\rangle)$, $(\langle c_5,c_4\rangle,\langle c_4,c_2\rangle)$, $(\langle c_5,c_4\rangle,\langle c_4,c_5\rangle)$, $(\langle c_5,c_4\rangle,\langle c_6,c_2\rangle)$, $(\langle c_5,c_4\rangle,\langle c_6,c_5\rangle)$;
- $(\langle c_5,c_5\rangle,\langle c_5,c_5\rangle)$, $(\langle c_5,c_5\rangle,\langle c_4,c_5\rangle)$, $(\langle c_5,c_5\rangle,\langle c_6,c_5\rangle)$, $(\langle c_5,c_5\rangle,\langle c_5,c_4\rangle)$, $(\langle c_5,c_5\rangle,\langle c_5,c_6\rangle)$, $(\langle c_5,c_5\rangle,\langle c_4,c_4\rangle)$, $(\langle c_5,c_5\rangle,\langle c_4,c_6\rangle)$, $(\langle c_5,c_5\rangle,\langle c_6,c_4\rangle)$, $(\langle c_5,c_5\rangle,\langle c_6,c_6\rangle)$;
- $(\langle c_5,c_6\rangle,\langle c_5,c_6\rangle)$, $(\langle c_5,c_6\rangle,\langle c_4,c_6\rangle)$, $(\langle c_5,c_6\rangle,\langle c_6,c_6\rangle)$, $(\langle c_5,c_6\rangle,\langle c_5,c_6\rangle)$, $(\langle c_5,c_6\rangle,\langle c_4,c_5\rangle)$, $(\langle c_5,c_6\rangle,\langle c_6,c_5\rangle)$;
- $(\langle c_6,c_1\rangle,\langle c_6,c_1\rangle)$, $(\langle c_6,c_1\rangle,\langle c_5,c_1\rangle)$, $(\langle c_6,c_1\rangle,\langle c_6,c_2\rangle)$, $(\langle c_6,c_1\rangle,\langle c_5,c_2\rangle)$;
- $(\langle c_6,c_2\rangle,\langle c_6,c_2\rangle)$, $(\langle c_6,c_2\rangle,\langle c_5,c_2\rangle)$, $(\langle c_6,c_2\rangle,\langle c_6,c_1\rangle)$, $(\langle c_6,c_2\rangle,\langle c_6,c_3\rangle)$, $(\langle c_6,c_2\rangle,\langle c_6,c_4\rangle)$, $(\langle c_6,c_2\rangle,\langle c_5,c_1\rangle)$, $(\langle c_6,c_2\rangle,\langle c_5,c_3\rangle)$, $(\langle c_6,c_2\rangle,\langle c_5,c_4\rangle)$;
- $(\langle c_6,c_3\rangle,\langle c_6,c_3\rangle)$, $(\langle c_6,c_3\rangle,\langle c_5,c_3\rangle)$, $(\langle c_6,c_3\rangle,\langle c_6,c_2\rangle)$, $(\langle c_6,c_3\rangle,\langle c_5,c_2\rangle)$;
- $(\langle c_6,c_4\rangle,\langle c_6,c_4\rangle)$, $(\langle c_6,c_4\rangle,\langle c_5,c_4\rangle)$, $(\langle c_6,c_4\rangle,\langle c_6,c_2\rangle)$, $(\langle c_6,c_4\rangle,\langle c_6,c_5\rangle)$, $(\langle c_6,c_4\rangle,\langle c_5,c_2\rangle)$, $(\langle c_6,c_4\rangle,\langle c_5,c_5\rangle)$;
- $(\langle c_6,c_5\rangle,\langle c_6,c_5\rangle)$, $(\langle c_6,c_5\rangle,\langle c_5,c_5\rangle)$, $(\langle c_6,c_5\rangle,\langle c_6,c_4\rangle)$, $(\langle c_6,c_5\rangle,\langle c_6,c_6\rangle)$, $(\langle c_6,c_5\rangle,\langle c_5,c_4\rangle)$, $(\langle c_6,c_5\rangle,\langle c_5,c_6\rangle)$;
- $(\langle c_6,c_6\rangle,\langle c_6,c_6\rangle)$, $(\langle c_6,c_6\rangle,\langle c_5,c_6\rangle)$, $(\langle c_6,c_6\rangle,\langle c_6,c_5\rangle)$, $(\langle c_6,c_6\rangle,\langle c_5,c_5\rangle)$.

• Comparing the outputs in the case of a single robot, the set of outputs now includes the observation $\{y_1, y_2\}$ (interpreted as $y_1 \wedge y_2$), since having two robots, one can be in c_4 activating y_1 and the other one in c_6 activating y_2. For example, if we denote by h_T the output function, then $h_T\left(\langle c_4, c_6\rangle\right) = \{y_1, y_2\}$, $h_T\left(\langle c_1, c_4\rangle\right) = \{y_1\}$ and $h_T\left(\langle c_5, c_5\rangle\right) = \emptyset$.

Notice that the number of nodes and the number of transitions of the transition system of the team increases exponentially with the number of robots. In particular, the number of discrete states is equal to $|C|^{|R|}$, where $|C|$ is the number of cells in the environment and $|R|$ the number of robots. In fact, this is the main drawback to this type of model. The transition system definition of a team of robots can be formally given as:

Definition 4.3 Given a team of n identical robots modeled by n transition systems $T_i = \langle C, c_{0i}, E, \overline{H}, h\rangle$, $i = 1, \ldots, |R|$, the transition system modeling the movement capabilities of the whole team of robots is $T_T = \left(C_T, c_{0T}, E_T, H, h_T\right)$, where:

- $C_T = C^{|R|}$ is the set of states ($C^{|R|}$ is the $|R|$-times Cartesian product of C with itself);
- $c_{0T} = \langle c_{01}, c_{02}, \ldots, c_{0|R|}\rangle \in C_T$ is the initial state;
- $E_T \subseteq C_T \times C_T$ is the transition relation, with $\left(\langle c_1, c_2, \ldots, c_n\rangle, \langle c_1', c_2', \ldots, c_{|R|}'\rangle\right) \in E_T$ if and only if $(c_i, c_i') \in E$, $\forall i = 1, \ldots, |R|$;
- $h_T : C_T \to 2^Y$ is the satisfaction function, with $h_T\left(\langle c_1, c_2, \ldots, c_{|R|}\rangle\right) = \bigcup_{i=1}^{|R|} h(c_i)$;
- $H = \underbrace{\overline{H} \times \overline{H} \times \cdots \times \overline{H}}_{|R|\text{-times}}$ is the set of possible observations, where $H \subseteq 2^Y$.

So far, the construction of T_T corresponds to a synchronous product of $|R|$ transition systems T_i, $i = 1, \ldots, |R|$, such as the one used in [112]. Note that here we neither restrict states of T_T such that at most one robot can be in a cell at a given state, and nor do we restrict transitions such that two robots cannot swap their cells. Such behaviors could lead to collisions in the case of robots with non-negligible size, but in such situations one can assume that local rules are used for avoiding collisions. Otherwise, restricting C_T and E_T such that collisions are avoided during the planning level would add additional conservatism to the solution, since T_T would be more restrictive. Instead of the power set 2^Y, the set H includes only those observations (tuples of regions) that can be generated by the team. Note that this model suffers from the state space explosion problem, since the number of states of the global transition system T_T grows exponentially with the number of robots in the environment. In order to overcome this state space explosion problem, Petri net models are introduced in the following section.

4.4 Petri net models

Petri nets is a modeling formalism for discrete event systems, especially useful for concurrent systems. The possible movements of a robot can be seen as a sequential system, while a team $R = \{r_1, r_2, \dots, r_{|R|}\}$ of $|R|$ robots may be seen as a parallel execution of $|R|$ sequential systems. Furthermore, the state in a Petri net is no longer global as in the case of transition systems (notice that the set C_T in Definition 4.3 is defined as a Cartesian product of $|R|$ sets C). The state in a Petri net is numeric (not symbolic as in transition system models) and distributed, allowing the definition of a state equation that can be used for analysis and design purposes.

Let us first intuitively define the Petri net (PN) model of the two robots evolving in the discretized environment in Figure 4.1. The PN model is shown in Figure 4.4. For each cell c_i of the partitioned environment a place p_i is defined, giving the set of all places $P = \{p_1, p_2, p_3, p_4, p_5, p_6\}$. For each pair of adjacent cells c_i and c_j, two transition nodes will be added, t_{ij} and t_{ji}. Consider for example the adjacent cells c_1 and c_2 modeled by places p_1 and p_2. Transitions t_{12} and t_{21} are added, modeling the possible movement of a robot from c_1 to c_2 or from c_2 to c_1, respectively. When transition t_{12} fires, a token will be removed from p_1 and a token will be created in p_2, representing the movement of a robot from c_1 to c_2. The set of all transitions corresponding to this environment is $T = \{t_{12}, t_{21}, t_{23}, t_{32}, t_{24}, t_{42}, t_{45}, t_{54}, t_{56}, t_{65}\}$.

As mentioned, the tokens represent the robots. Since initially there is one robot in c_1 and one robot in c_3, one token will be added to p_1 and one to p_3. The distribution of tokens between places is called the *marking* or the *state* of the PN system

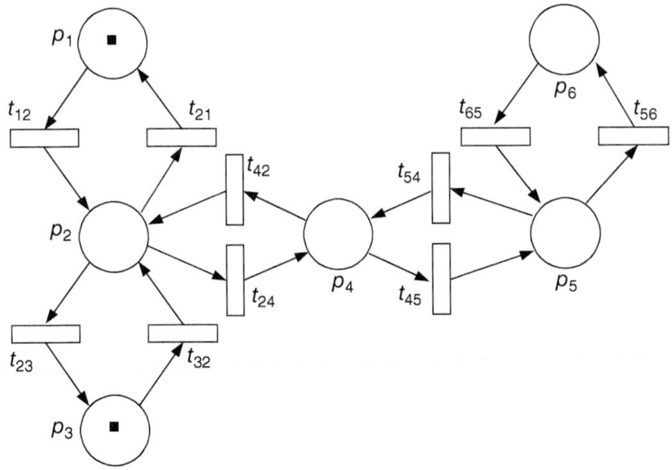

Figure 4.4 Example of a Petri net model of a team of robots. Petri net model of the robots in the discretized environment of Figure 4.1.

and is in general given as a vector. The initial marking is denoted by $\boldsymbol{m}_0$, being for the PN system in Figure 4.4 equal to $\boldsymbol{m}_0 = [1,0,1,0,0,0]^T$. Notice that it is impossible to differentiate between the tokens and for this reason the mobile robots are assumed to be identical. If one wants to model robots with different dynamics, color Petri nets [103] can be used, but this is beyond the scope of this book.

The formal definition of the PN is given in the following definition.

Definition 4.4 Given an environment abstracted to a graph $G = \langle C, E, O\rangle$ and a team of $|R|$ identical omnidirectional point robots, the Robot Motion Petri Net (RMPN) system model is a 4-tuple $\mathcal{Q} = \langle \mathcal{N}, \boldsymbol{m}_0, H, h\rangle$, where:

- $\mathcal{N} = \langle P, T, F\rangle$ is the Petri net structure with
  - P the finite set of places (one place p_i per each cell $c_i \in C$, i.e. $|P| = |C|$);
  - T the finite set of transitions modeling the movement capabilities of the robots between cells. For each edge $(c_j, c_k) \in E$ with $c_j \neq c_k$ two transitions denoted as t_{jk} and t_{kj} are included in T;
  - $F \subseteq (P \times T) \cup (T \times P)$ the set of arcs, such that for all $t_{jk}, t_{kj} \in T$ corresponding to the edge $(c_j, c_k) \in E$ the following four arcs are added: $F = F \cup \{(p_j, t_{jk}), (t_{jk}, p_k), (p_k, t_{kj}), (t_{kj}, p_j)\}$.
- $\boldsymbol{m}_0$ is the initial marking, where $\boldsymbol{m}_0[p_i]$ is the initial number of tokens in place p_i and is equal to the number of robots that initially exist in region c_i;
- $h : P \to 2^Y$ (with $Y = \{y_1, y_2, \dots, y_{|O|}\}$ is an output (or observation) function, where $h(p_i)$ is the conjunction of all outputs active in place p_i (corresponding to cell c_i). If p_i has at least one token, then observations from $h(p_i)$ are active;
- $H = \underbrace{\overline{H} \times \overline{H} \times \dots \times \overline{H}}_{|R|\text{-times}}$ is the set of possible outputs (observations) of the team, where $\overline{H} = \bigcup_{p \in P} h(p)$ is the set of output symbols generated by one robot. Again, since there are more robots and some regions of interest may overlap, an element of H can be seen as a conjunction of elements y_i. ■

Let $\boldsymbol{m}$ be a reachable marking of the net system, where $\boldsymbol{m} = \boldsymbol{m}_0$ is the initial marking. In order to get a state equation for the RMPN, the flow relation of the set F in Definition 4.4 can be equivalently represented by two incidence matrices $\boldsymbol{Pre} \in \{0,1\}^{|P|\times|T|}$ and $\boldsymbol{Post} \in \{0,1\}^{|P|\times|T|}$, defined as follows:

$$\boldsymbol{Pre}[i,j] = \begin{cases} 1, & \text{if } (p_i, t_j) \in F, \\ 0, & \text{otherwise.} \end{cases} \quad \boldsymbol{Post}[i,j] = \begin{cases} 1, & \text{if } (t_j, p_i) \in F, \\ 0, & \text{otherwise.} \end{cases} \tag{4.3}$$

For $p \in P$, the sets of its input and output transitions are denoted as ${}^\bullet p = \{t \in T | \boldsymbol{Post}[p,t] > 0\}$ and $p^\bullet = \{t \in T | \boldsymbol{Pre}[p,t] > 0\}$, respectively. In a similar manner, for a transition $t \in T$, there are defined the sets of its input and output places, denoted by ${}^\bullet t$ and $t^\bullet$, respectively A transition $t_j \in T$ is enabled at a given reachable marking $\boldsymbol{m}$ if all its input places contain at least one token, i.e. $\forall p_i \in {}^\bullet t_j, \boldsymbol{m}[p_i] \geq 1$. An enabled transition t_j can fire, leading to a new state

$$\tilde{\boldsymbol{m}} = \boldsymbol{m} + \boldsymbol{C}[\cdot, t_j],$$

where $\boldsymbol{C} = \boldsymbol{Post} - \boldsymbol{Pre}$ is the token flow matrix and $\boldsymbol{C}[\cdot, t_j]$ is the column corresponding to t_j. It will be said that $\tilde{\boldsymbol{m}}$ is a reachable marking that has been reached from $\boldsymbol{m}$ by firing t_j and is written as $\boldsymbol{m}[t_j\rangle\tilde{\boldsymbol{m}}$.

In an RMPN, the firing of a transition t corresponds to the movement of a robot from cell ${}^\bullet t = \{p_i\}$ to cell $t^\bullet = \{p_j\}$. Notice that by definition, each transition has only one input and only one output place, thus leading to a *state machine* PN [158]. If $\tilde{\boldsymbol{m}}$ is reachable from $\boldsymbol{m}$ through a finite sequence of transitions $\sigma = t_{i_1} t_{i_2} \cdots t_{i_k}$, the following state (or fundamental) equation is satisfied:

$$\tilde{\boldsymbol{m}} = \boldsymbol{m} + \boldsymbol{C} \cdot \sigma, \tag{4.4}$$

where $\sigma \in \mathbb{N}_{\geq 0}^{|T|}$ is the firing count vector, i.e., its jth element is the cumulative amount of firings of t_j in the sequence σ. Notice that Eq. (4.4) is only a necessary condition for the reachability of a marking. The marking solutions of (4.4) that are not reachable are called *spurious markings*. In general, checking if a marking $\boldsymbol{m}$ is reachable or not is not an easy problem due to these spurious markings [192].

Example 4.1 *Let us consider the PN system in Figure 4.5:*

- *The set of places* $P = \{p_1, p_2, p_3\}$*;*
- *The set of transitions* $T = \{t_1, t_2\}$*;*
- *The set of arcs* $F = \{(p_1, t_1), (t_1, p_2), (t_1, p_3), (p_2, t_2), (t_2, p_1)\}$*;*
- *The initial marking* $\boldsymbol{m}_0 = [0, 0, 0]^T$.

The pre and post incidence matrices are:

$$\boldsymbol{Pre} = \begin{bmatrix} 1 & 0 \\ 0 & 1 \\ 0 & 0 \end{bmatrix} \text{ and } \boldsymbol{Post} = \begin{bmatrix} 0 & 1 \\ 1 & 0 \\ 1 & 0 \end{bmatrix}. \tag{4.5}$$

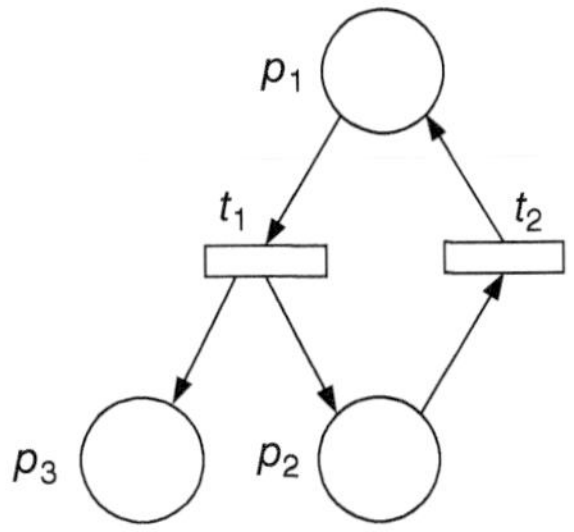

Figure 4.5 Spurious markings in a Petri net system. A small Petri net system with $\boldsymbol{m}_0 = [0, 0, 0]^T$ for which $\boldsymbol{m} = [0, 0, 1]^T$ is a spurious marking.

Therefore,

$$\boldsymbol{C} = \boldsymbol{Post} - \boldsymbol{Pre} = \begin{bmatrix} -1 & 1 \\ 1 & -1 \\ 1 & 0 \end{bmatrix}$$

and it is easy to observe that $\tilde{\boldsymbol{m}} = [0, 0, 1]^T$ *is a spurious marking. Indeed, if* $\sigma = [1, 1]^T$ *and* $\boldsymbol{m} = \boldsymbol{m}_0$*, according to equation (4.4) we get*

$$\boldsymbol{m}_0 + \boldsymbol{C} \cdot \sigma = \begin{bmatrix} 0 \\ 0 \\ 0 \end{bmatrix} + \begin{bmatrix} -1 & 1 \\ 1 & -1 \\ 1 & 0 \end{bmatrix} \cdot \begin{bmatrix} 1 \\ 1 \end{bmatrix} = \begin{bmatrix} 0 \\ 0 \\ 1 \end{bmatrix} = \tilde{\boldsymbol{m}}.$$

Hence, $\tilde{\boldsymbol{m}}$ *is a solution of the state-equation (4.4), but it is not reachable in the PN system of Figure 4.5. In fact, the PN system is in a deadlock state for the marking in the figure since no transition is enabled. The firing sequence* $\sigma = [1, 1]^T$ *solution of (4.4) does not correspond to any firing sequence since neither* t_1 *or* t_2 *can fire.*

Let us assume now the initial marking is $\boldsymbol{m}'_0 = [1, 0, 0]^T$*. The marking* $\boldsymbol{m}'$ *that is reached after the firing of the sequence* $\sigma' = t_1t_2t_1t_2t_1$ *is obtained from (4.4), with* $\sigma' = [3, 2]^T$ *(the firing vector corresponding to the firing sequence* σ*)*

$$\boldsymbol{m}' = \boldsymbol{m}'_0 + \boldsymbol{C} \cdot \sigma' = \begin{bmatrix} 1 \\ 0 \\ 0 \end{bmatrix} + \begin{bmatrix} -1 & 1 \\ 1 & -1 \\ 1 & 0 \end{bmatrix} \cdot \begin{bmatrix} 3 \\ 2 \end{bmatrix} = \begin{bmatrix} 0 \\ 1 \\ 2 \end{bmatrix}.$$

A PN is live if from any reachable marking any transition can eventually fire (possibly after first firing other transitions). It is well known that for state machine PNs, liveness is equivalent to strong connectedness (a PN is strongly connected if there exists a path from any node to any other node of the PN) and non-emptiness of (initial) marking. Moreover, in a live state machine, there exist no spurious markings [192], i.e. the solutions of the fundamental equation (4.4) give the set of reachable markings.

Let $\boldsymbol{v}_i \in \{0, 1\}^{1\times|P|}$ be the characteristic row vector of the proposition y_i such that $\boldsymbol{v}_i[p_k] = 1$ if $y_i \in h(p_k)$ and $\boldsymbol{v}_i[p_k] = 0$ otherwise. It is easy to observe that, for a reachable marking $\boldsymbol{m}$, if the product $\boldsymbol{v}_i \cdot \boldsymbol{m} > 0$ then the output y_i is active at $\boldsymbol{m}$.

Let $\boldsymbol{V} \in \{0, 1\}^{|O|\times|P|}$ be the matrix formed by the characteristic vectors of all observations, that is, the ith row is the characteristic vector of y_i, $i = 1, \ldots |O|$. The product $\boldsymbol{V} \cdot \boldsymbol{m}$ is a column vector of dimension $|O|$ where the ith element is non-zero if observation y_i is active. We denote by $||\boldsymbol{V} \cdot \boldsymbol{m}||$ the set of outputs corresponding to non-zero elements of $\boldsymbol{V} \cdot \boldsymbol{m}$, that is, $||\boldsymbol{V} \cdot \boldsymbol{m}||$ is the set of active observations at marking $\boldsymbol{m}$.

Example 4.2 *Let us consider again the PN system given in Figure 4.4 with the initial marking in the figure*

$$\boldsymbol{m}_0 = \left[\boldsymbol{m}_0[p_1], \boldsymbol{m}_0[p_2], \boldsymbol{m}_0[p_3], \boldsymbol{m}_0[p_4], \boldsymbol{m}_0[p_5], \boldsymbol{m}_0[p_6]\right]^T = [1,0,1,0,0,0]^T. \tag{4.6}$$

Assume that the order of the transition firings in the firing count vector is the following:

$$\sigma = \left[\sigma[t_{12}], \sigma[t_{21}], \sigma[t_{23}], \sigma[t_{32}], \sigma[t_{24}], \sigma[t_{42}], \sigma[t_{45}], \sigma[t_{54}], \sigma[t_{56}], \sigma[t_{65}]\right]^T$$

while the output of place p_4 is y_1, giving the characteristic vector of y_1 as

$$\boldsymbol{v}_1 = [0,0,0,1,0,0]. \tag{4.7}$$

Assume that both robots should arrive in p_4. Therefore, the desired final marking (state) is $\boldsymbol{m}_d = [0,0,0,2,0,0]^T$. The firing count vector to reach $\boldsymbol{m}_d$ can be obtained by solving the following Mixed Integer Linear Programming (MILP) problem

$$\begin{aligned} \min\ & \boldsymbol{1} \cdot \sigma \\ \text{s.t. } & \boldsymbol{m} = \boldsymbol{m}_0 + \boldsymbol{C} \cdot \sigma \\ & \boldsymbol{v}_1 \cdot \boldsymbol{m} = 2 \\ & \boldsymbol{m} \in \mathbb{N}_{\geq 0}^{|P|\times 1}, \sigma \in \mathbb{N}_{\geq \boldsymbol{0}}^{|T|\times 1} \end{aligned} \tag{4.8}$$

where $\boldsymbol{m}_0$ is given in (4.6), $\boldsymbol{v}_1$ is given in (4.7), while $\boldsymbol{C}$ is the corresponding incidence matrix,

$$\boldsymbol{C} = \begin{array}{c} \\ p_1 \\ p_2 \\ p_3 \\ p_4 \\ p_5 \\ p_6 \end{array} \begin{array}{c} \begin{array}{cccccccccc} t_{12} & t_{21} & t_{23} & t_{32} & t_{24} & t_{42} & t_{45} & t_{54} & t_{56} & t_{65} \end{array} \\ \left[\begin{array}{cccccccccc} -1 & 1 & 0 & 0 & 0 & 0 & 0 & 0 & 0 & 0 \\ 1 & -1 & -1 & 1 & -1 & 1 & 0 & 0 & 0 & 0 \\ 0 & 0 & 1 & -1 & 0 & 0 & 0 & 0 & 0 & 0 \\ 0 & 0 & 0 & 0 & 1 & -1 & -1 & 1 & 0 & 0 \\ 0 & 0 & 0 & 0 & 0 & 0 & 1 & -1 & -1 & 1 \\ 0 & 0 & 0 & 0 & 0 & 0 & 0 & 0 & 1 & -1 \end{array}\right] \end{array} \tag{4.9}$$

It is easy to observe that $\boldsymbol{m}_d = [0,0,0,2,0,0]^T$ is the solution of MILP (4.8), with $\sigma = [1,0,0,1,2,\ 0,0,0,0,0]^T$. This firing count vector corresponds to a firing sequence in which t_{12} and t_{32} should fire once while t_{24} should fire twice. For this particular case, the firing sequence is easy to obtain (by firing transitions as soon as they are enabled), one such firing sequence being, for example, $\sigma = t_{12}t_{24}t_{32}t_{24}$. Notice that in this case, the optimal paths (optimal with respect to the number of transition firings) do not contain any cycles; thus the robots reach p_4 without going through unnecessary cells.

Furthermore, since the RMPN is a state machine, MILP of Eq. (4.8) can be relaxed to a linear programming problem (with the same constraints and cost function), since the solution is an integer [115]. Therefore, the complexity of this problem is linear in the computing time.

Let us assume now that one robot should reach first p_6 and then go to p_4 while the second robot should go directly to p_4. The first idea is to use MILP of Eq. (4.8) by adding the new constraint $\sigma[t_{56}] \geq 1$. Notice that if transition t_{56} fires then a robot is entering in p_6. Therefore, the new MILP is

$$\begin{aligned} \min\ & \mathbf{1} \cdot \sigma \\ \text{s.t. } & \boldsymbol{m} = \boldsymbol{m}_0 + \boldsymbol{C} \cdot \sigma \\ & \boldsymbol{m} \cdot \boldsymbol{v}_1 = 2 \\ & \sigma[t_{56}] \geq 1 \\ & \boldsymbol{m} \in \mathbb{N}_{\geq 0}^{|P|\times 1}, \sigma \in \mathbb{N}_{\geq \boldsymbol{0}}^{|T|\times 1} \end{aligned} \tag{4.10}$$

where $\boldsymbol{m}_0$ and $\boldsymbol{v}_1$ have the same values as in MILP of Eq. (4.8). Solving MILP of Eq. (4.10), the solution is $\boldsymbol{m} = [0, 0, 0, 2, 0, 0]^T$ (the desired final marking) and $\sigma = [1, 0, 0, 1, 2, 0, 0, 0, 1, 1]^T$. Notice that this firing vector does not correspond to any firing sequence and so we will call it a spurious firing vector. *This happens because now one robot is required to perform two cycles after stopping in p_4. In particular, $t_{56}t_{65}$ and $t_{45}t_{54}$ correspond to the firing sequence: $t_{45}t_{56}t_{65}t_{54}$. Since in a state machine PN, each cycle is a T-semiflow (by firing it the marking does not change), when solving the MILP of Eq. (4.10) some cycles will not be included in the firing vector. In this case, the firing of the cycle $t_{45}t_{54}$ is not included since by firing it the marking does not change and the constraints are satisfied. From $\sigma = [1, 0, 0, 1, 2, 0, 0, 0, 1, 1]^T$ it is not possible to obtain any firing sequence.*

One possibility is to consider an intermediate marking $\boldsymbol{m}_i$ such that from $\boldsymbol{m}_0$ to $\boldsymbol{m}_i$ and from $\boldsymbol{m}_i$ to $\boldsymbol{m}_d$ no cycle will be required. For this, let us define a new output y_2 that will be assigned to p_6. The characteristic vector of y_2 is

$$\boldsymbol{v}_2 = [0, 0, 0, 0, 0, 1]. \tag{4.11}$$

Intermediate marking $\boldsymbol{m}_i$ will correspond to at least one robot in p_6. The new MILP is

$$\begin{aligned} \min\ & \mathbf{1} \cdot \sigma \\ \text{s.t. } & \boldsymbol{m}_i = \boldsymbol{m}_0 + \boldsymbol{C} \cdot \sigma_i \\ & \boldsymbol{m}_i \cdot \boldsymbol{v}_2 \geq 1 \\ & \boldsymbol{m} = \boldsymbol{m}_i + \boldsymbol{C} \cdot \sigma \\ & \boldsymbol{m} \cdot \boldsymbol{v}_1 = 2 \\ & \boldsymbol{m}, \boldsymbol{m}_i \in \mathbb{N}_{\geq 0}^{|P|\times 1}, \sigma, \sigma_i \in \mathbb{N}_{\geq \boldsymbol{0}}^{|T|\times 1} \end{aligned} \tag{4.12}$$

Given a graph abstraction of an environment, the informal steps that lead to construct the RMPN model are captured in Algorithm 4.2. The transitions added on line 8 assume fully actuated point robots, which can move from the current cell to any adjacent cell by straight movement to the middle point of the line segment shared by the two cells. Algorithm 4.2 also returns the vector $\boldsymbol{w} \in \mathbb{R}_{\geq 0}^{|T|}$ that contains the average distance for traveling between adjacent cells. Due to the convex polygonal cells and the piece-wise straight movements of robots, an expected distance for moving from cell c_i to c_j is chosen in line 10 as being the average of distances between the exit point (middle of the line segment shared by c_i and c_j) and any possible entry point in c_i (middle of line segments shared by c_i with all neighboring cells different from c_j). Similarly, the expected distance for moving from c_j to c_i is chosen in line 11. For different robot dynamics, the condition from line 7 can be replaced with the existence of control laws steering the robot from cell c_i (c_j) to the adjacent cell c_j (c_i) in finite time; e.g. works such as [12, 14, 91] describe the case of affine or multi-affine dynamics in polytopal or rectangular environments. Lines 4 and 12 add the tokens, based on robots' initial positions. The observation map in line 15 is well-defined, since the used cell decomposition techniques for regions of interest preserve boundaries and intersections of regions from O, and therefore all points inside a cell satisfy the same set of regions.

Algorithm 4.2: Construct the RMPN system Q

Input: Graph abstraction of the environment $G = \langle C, E, O \rangle$ and adjacency matrix A

Output: Team model Q

1 Associate each cell c_i from decomposition to a place $p_i \in P$. Let $P = \{p_1, p_2, \dots, p_{|C|}\}$;
2 Let $T = \emptyset$, $F = \emptyset$, $\boldsymbol{w} = \boldsymbol{0}$;
3 Let $h(p_i) = \emptyset, \forall p_i \in P$;
4 **for** $i = 1$ *to* $|C| - 1$ **do**
5     $\boldsymbol{m}_0[p_i]$ = number of robots initially deployed in cell c_i;
6   **for** $j = i + 1$ *to* $|C|$ **do**
7   **if** $(A[i, j] == 1)$ **then**
8             $T = T \cup \{t_{i,j}, t_{j,i}\}$;
9             $F = F \cup \{(p_i, t_{i,j}), (t_{i,j}, p_j), (p_j, t_{j,i}), (t_{j,i}, p_i)\}$;
10             $\boldsymbol{w}[t_{i,j}]$ = average distance from cell c_i to c_j;
11             $\boldsymbol{w}[t_{j,i}]$ = average distance from cell c_j to c_i;

12 $\boldsymbol{m}_0[p_{|C|}]$ = number of robots initially deployed in cell $c_{|C|}$;
13 **forall** $O_i \in O$ **do**
14   **forall** $c_j \in O_i$ **do**
15         $h(p_j) = h(p_j) \cup \{y_i\}$;

Table 4.1 Comparison between the discrete formalisms used in multi-robot path planning, where C is the set of discrete cells obtained after cell decomposition and E is the set of edges computed based on the adjacency relation.

	Modeling		Planning/Synthesis	
	# Discrete nodes	**Scalable**	**Method**	**Solution**
Transition systems	$\lvert C\rvert^{\#\text{ of robots}}$	no	search on the graph	path
Petri nets	$\lvert C\rvert + \lvert E\rvert$	yes	mathematical programming	firing vector

Matlab function from RMTool: *rmt_construct_PN*. This functions allows the user to obtain the PN model of a partitioned environment.

Syntax: $[Pre, Post] = rmt\_construct\_PN(Adj)$

- *Adj* is the adjacency matrix of the environment, as defined in Eq. (4.1);
- *Pre* and *Post* are Matlab matrices of the corresponding incidence matrices.

The corresponding incidence matrices and the initial marking of a given environment can be exported to Matlab workspace for future analysis by using the option *Environment* → *Export to workspace* option from the RMTool menu.

Table 4.1 presents a comparison of both discrete formalisms used to solve the path planning problem. The number of nodes of the team transition systems grows exponentially with the number of robots. However, the complexity of the path planning algorithms based on graph searches is polynomial with the number of nodes. Conversely, the number of nodes of the PN models is equal to the number of cells plus the number of edges and hence are scalable with respect to the number of robots (this does not depend on the number of robots). However, the path planning algorithms are based on solving MILP problems and the main difficulty is that the solution of the optimization problem is a firing vector from which in some cases it is difficult to get firing sequences (to obtain the robot path). By simulations that we have done, we concluded that for one robot or teams of 2 or 3 robots, it may be better to use transition system models while for more robots the usage of PN models is feasible.

4.5 Petri nets in resource allocation systems models

In this section, we consider that some regions of the environment have finite capacity, that is, a finite number of robots can be located inside such regions at any time moment. This constraint can be used by a team of robots to avoid collision between them. In the most restrictive case, by imposing that in each region only one robot can exist at every time moment, the collisions cannot occur (it is not possible for two robots to swap their cells, because for a brief moment there would be two robots on the common cell border, and thus in both cells). However, these capacity constraints introduce waiting modes in the robot motion that may lead to another problem: the deadlock. The PN model that will be presented in this section can be used together with the existing theoretical results from PN literature to compute controllers in order to prevent the occurrence of deadlocks (see Section 6.6).

Since the deadlock prevention algorithms are computationally very hard, the team model will not include all possible movements of the robots given by the graph abstraction in Definition 4.1. Given a number of $|R|$ identical robots, we assume that each robot, i, may choose one path from a given set $\mathcal{T}_i$. All paths from $\mathcal{T}_i$ start and finish in the same region c_{0i}.

For example, consider the planar environment partitioned in 27 cells in Figure 4.6 in which five regions of interest exist, $O_1 = \{c_{18}\}$, $O_2 = \{c_{23}\}$, $O_3 = \{c_{16}\}$, $O_4 = \{c_2\}$, and $O_5 = \{c_{27}\}$. Moreover, assume that there are two robots, both initially placed in region c_{25} (represented by a circle and a square). The mission requires that both robots should enter simultaneously to c_{27} and c_{16}, then the robot that enters in c_{27} should go to c_2 and go back to region c_{25}, while the robot that enters in c_{16} should come back to c_{25}. There are several possible paths for each robot, but for simplicity of exposition, we consider only one path for each robot, as follows:

- Robot r_1 goes through regions: $c_{25}, c_{22}, c_{27}, c_{22}, c_9, c_{20}, c_{24}, c_1, c_8, c_2, c_8, c_1, c_{24}, c_{20}, c_9, c_{22}, c_{25}$;
- Robot r_2 goes through regions: $c_{25}, c_{22}, c_9, c_{20}, c_{24}, c_{15}, c_{10}, c_{16}, c_{11}, c_1, c_{24}, c_{20}, c_9, c_{22}, c_{25}$.

Assume that in order to avoid collisions, the environment regions are restricted such that in all regions except c_{25} at most one robot can be at any time moment. Obviously, if the robots individually follow their paths without any additional rule, the region capacities may be violated (e.g. in c_{22}, c_9, c_1, c_{20}, or c_{24}). Even if local waiting strategies are used for avoiding this, deadlocks may occur (e.g. if r_1 is in c_{20} trying to go to c_{24}, while r_2 is in c_{24} trying to go to c_{20}).

The PN model constructed in this section accounts only for region capacity, while the deadlock-prevention strategies will be added in Section 6.6. The PN corresponding to the above example is shown in Figure 4.7 and its structure is explained next. Not all arcs are represented in the figure for sake of clarity. In particular, input and output arcs in the following places are drawn only partially: c_1, c_2, c_8, c_9, c_{15}, c_{20}, c_{24}, and c_{27}.

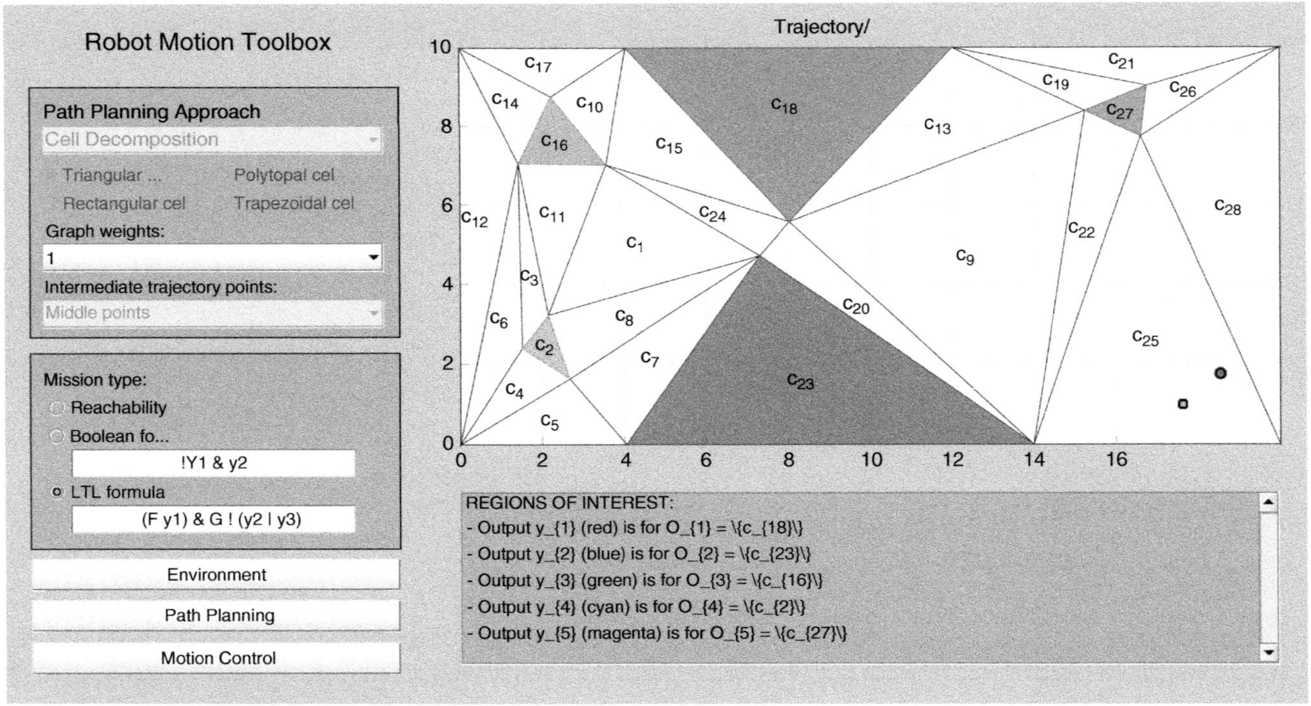

Figure 4.6 Example of a discretized environment and two robots. A discretized environment defined in RMTool containing with five regions of interest and two robots.

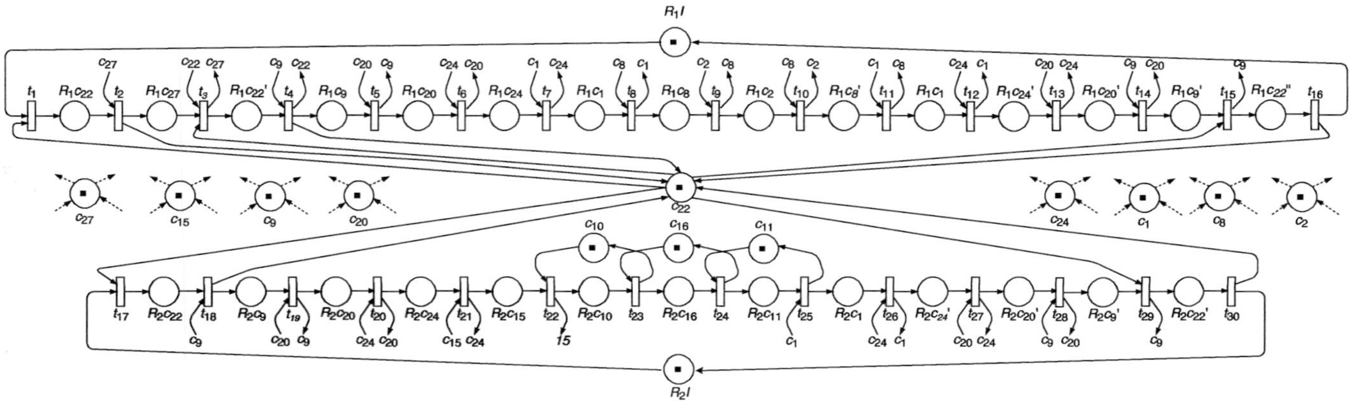

Figure 4.7 Example of an RAS PN model. The RAS PN model of two robots evolving in the environment given in Figure 4.6 that should follow two different paths.

Each successive region of the paths is modeled by a place. For example, the place R_1c_{22} corresponds to the presence of the first robot in region c_{22}. The firing of its input transition t_1 will model the entrance of robot r_1 in region c_{22} while the firing of its output transition t_2 will model the exit of robot r_1 from c_{22} and the entrance in the following one, i.e. c_{27}. For each restricted region (here we assume capacity one but a finite capacity greater than one may be considered) a new place is introduced modeling the limited capacity. For example, place c_{22} corresponds to the capacity of region c_{22}, and it is initially marked with a number of tokens equal to the capacity of c_{22} (here equal to one).

Places R_1c_{22}, R_1c_{22}', R_1c_{22}", R_2c_{22}, and R_2c_{22}' correspond to robots visiting region c_{22}, and they are connected via transitions to c_{22} such that only one robot can be at a time instant in the corresponding region. Assume that, initially, when both robots are traveling toward c_{22}, robot r_1 arrives first and enters this restricted region. Then, one token is removed from the place corresponding to its capacity (c_{22}). Thus, even if r_2 arrives to c_{22} while r_1 is still there, it cannot enter the region because transition t_{17} cannot be fired without one token in c_{22}. The token in c_{22} is returned (released) when r_1 exits c_{22}, and then r_2 can enter c_{22}. In other words, c_{22} can be viewed as a shared resource for robots r_1 and r_2, and the marking of c_{22} with its input and output transition models the correct access to this resource. The same idea applies to more robots and for restricted capacities larger than one, by simply placing the corresponding number of tokens in the place modeling each capacity.

All places that correspond to regions that appear along a path are connected by transitions in their appearing order, e.g. R_1c_{22} is connected to R_1c_{27} via t_2 (meaning that r_1 crosses from c_{22} to c_{27}) and so on. Note that if a region is visited more than once along the same path, more places are added to the PN model. For example, there are six places corresponding to c_{22}: one models its limited capacity (c_{22}), three model its occurrence along path of r_1 (R_1c_{22}, R_1c_{22}', and R_1c_{22}"), and two model its occurrence along path of r_2 (R_2c_{22} and R_2c_{22}').

A number of $|R|$ places model the *Idle state* of the robots: R_1I and R_2I. One token in R_1I (or R_2I) means that the robot 1 (or 2) is either waiting to begin its motion or it stopped after finishing the path. If there are more possible paths for a robot i ($|\mathcal{T}_i| > 1$), its idle place will have $|\mathcal{T}_i|$ output transitions, one for each path. Choosing one of those paths means firing the corresponding transition and moving the token from the idle state to a state corresponding to actual movement.

Remark 4.1 Note that when a robot finishes its chosen path, the token is returned to the robot's idle place. However, the PN model could be used again for supervising the motion only after the robot is placed back in its initial region of the environment. Thus, when tokens are returned to all idle places, the mission is accomplished. We do not model the mission accomplishment with different places (we do not have places modeling where the mission has been finished) from the idle ones, because tokens arrived in such places would remain there forever, and the PN would belong to a different class (for which deadlock avoidance strategies might not be characterized).

As a summary of the PN model constructed so far, some places correspond to the regions visited by the robots along their paths, other places correspond to limited capacity of regions, and $|R|$ idle places correspond to beginning or ending the assignment. The transitions of the PN correspond to robots entering new regions from their paths, and the connections between places and transitions ensure that the capacity of each visited region is satisfied.

The formal definition and construction procedure for the PN model corresponding to robot motions is similar to the definition of the well-known class of Petri nets used in Resource Allocation Systems, called $L - S^3PR$ from [64].

Definition 4.5 The Resource Allocation Robot Motion PN (RARMPN) is a PN $\mathcal{N} = \langle P, T, F \rangle$ such that:

1) $P = P_T \cup P_C \cup P_I$ is a partition such that:
   a) $P_I = \left\{p_I^1, \dots, p_I^{|R|}\right\}$, $|R| > 0$ ($p_I^i \in P_I$ is called the idle place of robot r_i);
   b) $P_T = \bigcup_{i=1}^{|R|} P_T^i$, where $P_T^i \cap P_T^j = \emptyset$, for all $i \neq j$ ($p \in P_T^i$ is modeling the presence of the robot r_i in a region from one of its possible paths);
   c) $P_C = \{c_i | c_i \text{ is a region of at least one path}\}$.
2) $T = \bigcup_{i=1}^{|R|} T^i$, where $T^i \cap T^j = \emptyset$, for all $i \neq j$.
3) $\forall i \in \{1, \dots, |R|\}$, the subnet $\mathcal{N}^i$ generated by $\left\{p_I^i\right\} \cup P_T^i \cup T^i$, is a strongly connected state machine such that every cycle contains $\left\{p_I^i\right\}$ and $\forall p \in P_T^i$, $|p^\bullet| = 1$.
4) $\mathcal{N}$ is strongly connected.
5) $\forall i \in \{1, \dots, k\}$, $\forall p \in P_T^i$, ${}^{\bullet\bullet}p \cap P_C = p^{\bullet\bullet} \cap P_C$, and $|{}^{\bullet\bullet}p \cap P_C| = 1$.

Based on the informal explanations from the beginning of this section, the set of places of RARMPN from Definition 4.5 is composed of: the set P_I of idle places (denoted by P_0 in the standard definition of $L - S^3PR$), the set P_T of path places (P_A in the definition of $L - S^3PR$)), and the set P_C modeling region capacities (set P_R in the definition of $L - S^3PR$). The subnet $\mathcal{N}^i$ corresponds to all possible paths of robot i. A cycle in this subnet is a single path from set $\mathcal{T}_i$, which guarantees the last equality from item 3 in Definition 4.5. Item 5 of Definition 4.5 ensures that all regions among paths are with limited capacity. Notice that this is not a real constraint since it is always possible to impose a large capacity, e.g. greater than the number of robots.

An initial admissible marking of RARMPN is given by the following definition.

Definition 4.6 Let $\mathcal{N} = \langle P_T \cup P_C \cup P_I, T, F\rangle$ be a RARMPN. An initial marking $\boldsymbol{m}_0$ is called admissible if:

- $\boldsymbol{m}_0[p_I^i] = 1, \forall p_I^i \in P_I$;
- $\boldsymbol{m}_0[p_c] = cap(c_i) \geq 1, \ \forall p_c \in P_C$, where p_c is place modeling the capacity of region c_i and $cap(c_i)$ is the capacity of region c_i;
- $\boldsymbol{m}_0[p] = 0, \forall p \in P_T$.

In the RARMPN in Figure 4.7,

- $P_I = \{R_1I, R_2I\}$,
- $P_T^1 = \{R_1c_{22}, R_1c_{27}, R_1c_{22}', R_1c_9, R_1c_{20}, R_1c_{24}, R_1c_1, R_1c_8, R_1c_2, R_1c_8', R_1c_1', R_1c_{24}', R_1c_{20}', R_1c_9' R_1c_{22}''\}$,
- $P_T^2 = \{R_2.c_{22}, R_2c_9, R_2c_{20}, R_2c_{24}, R_2c_{15}, R_2c_{10}, R_2c_{16}, R_2c_{11}, R_2c_1, R_2c_{24}', R_2c_{20}', R_2c_9', R_2c_{22}'\}$,
- $P_C = \{c_1, c_2, c_8, c_9, c_{10}, c_{11}, c_{15}, c_{16}, c_{20}, c_{22}, c_{24}, c_{27}\}$.

Given the environment graph G, the set of robots R, and for each robot r_i, $i = 1, \ldots, |R|$, the set $\mathcal{T}_i$ of its possible paths, Algorithm 4.3 can be used for obtaining the RARMPN model. Steps of Algorithm 4.3 follows the intuitive ideas presented: each path of each robot is modeled by a state machine (lines 7-10) and the state machine is connected to the robot's idle place (line 11). Then, the capacity of each encountered region is modeled by a place connected to all transitions that correspond to entering and exiting that region (lines 12-18). The initial marking of the RARMPN model contains one token in each idle place (line 4) and a corresponding number of tokens in each place modeling the capacity of restricted regions (lines 14 and 17). In Algorithm 4.3, for a place p that models the presence of a robot in a region, the corresponding location from C is denoted by c_p.

4.6 High-level specifications

The robot motion tasks will be defined on the set Y, whose elements correspond to regions of interest O, as mentioned at the beginning of Section 4.3. The problem to be solved is to automatically generate (if possible) control strategies ensuring the satisfaction of the given specification. Thus, we are facing the problem of controlling either a transition system T_S (Definition 4.2), a transition system T_T (Definition 4.3), or a Petri net system Q (Definition 4.4). We emphasize that we are interested in an automatic design of control strategies, rather than a manual switching between several predefined operating modes. The latter methodology, although not very complex computationally, would require a human supervisor who deeply understands both the task and the functioning of the system.

Algorithm 4.3: Obtain the RARMPN model for capacity constraints

Input: $G = \langle C, E, O\rangle, R, \mathcal{T}_1, \ldots, \mathcal{T}_{|R|}$; `/* environment abstraction, robots and all possible paths */`

Output: $\langle \mathcal{N}, \boldsymbol{m}_0\rangle$; `/* the RARMPN system */`

Set $P_C = \emptyset$

forall $r_i \in R$ **do**

  Add a place $P_I^i = \{p_I^i\}$ modeling the Idle state of r_i;

  Set initial marking of p_I^i equal to 1;

  Set $P_T^i = T^i = F^i = \emptyset$

  forall $tr_j \in \mathcal{T}_i$; `/* assume` $tr_j = c_1^j\ c_2^j\ \ldots\ c_{|tr_j|}^j$ `*/`

  do

    $\overline{P} = \{p_1^j, p_2^j, \ldots p_{|tr_j|}^j\}$;

    $\overline{T} = \{t_1^j, t_2^j, \ldots, t_{|tr_j|}^j\}$;

    $\overline{F} = \{(t_1^j, p_1^j), (t_2^j, p_2^j), \ldots, (t_{|tr_j|}^j, p_{|tr_j|}^j)\} \cup \{(p_1^j, t_2^j), (p_2^j, t_3^j), \ldots, (p_{|tr_j|-1}^j, t_{|tr_j|}^j)\}$;

    Add one arc from p_I^i to t_1^j and one arc from $t_{|tr_j|}^j$ to p_I^i

    forall $p \in \overline{P}$ **do**

      if *capacity of region* c_p *is modeled by* $p_c \in P_C$ **then**

        Add one arc from p_c to ${}^\bullet p$ and one arc from $p^\bullet$ to p_c

      else

        Add place p_c to P_C for modeling capacity of c_p

        Set initial marking of p_c equal to $cap(c_p)$

        Add one arc from p_c to ${}^\bullet p$ and one arc from $p^\bullet$ to p_c

    $P_T^i = P_T^i \cup \overline{P}$

    $T^i = T^i \cup \overline{T}$

    $F^i = F^i \cup \overline{F}$

  Set initial markings of places P_T^i as zero

Construct $\mathcal{N}$ using places $\left(\bigcup_{i=1}^{|R|} P_I^i\right) \cup \left(\bigcup_{i=1}^{|R|} P_T^i\right) \cup P_C$ and transitions $\bigcup_{i=1}^{|R|} T^i$

In this section we give an overview of some formalisms that can be used for specifying robot motion tasks, namely:

- regular expressions [27, 129],
- ω-regular expressions [129],
- Boolean-based formulas [148],
- temporal logics (LTL, CTL, and CTL*) [10, 37, 109].

For each of them, the expressivity power and the complexity of solving the controller design problem is briefly described.

A. *Regular expressions* [27, 129] are convenient short descriptions of regular languages, which are sets of finite sequences accepted by finite automaton. Therefore, regular expressions over $Y = \{y_1, y_2, \dots, y_{|Y|}\}$ can be used for specifying terminating robot behaviors. For example,

- *visiting some regions in a desired order* ("visit y_1 and then y_2", expressed by $(y_2 + y_3 + \cdots + y_{|Y|})^* \cdot y_1 \cdot (y_1 + y_3 + \cdots + y_{|Y|})^* \cdot y_2$, where the symbol "*" denotes the Kleene closure of a set (set of all finite length strings consisting of elements of that set) [27], "+" stands for disjunction (in this case union of sets), and "·" stands for concatenation),
- *following a certain pattern* ("from y_1 go to y_2 and then to y_3, while avoiding all other regions", expressed by $y_1 \cdot y_2 \cdot y_3$).

Although this formalism is close to natural language, specifying more complicated requirements can become very hard. For a finite transition system T_S, the problem of generating a control strategy from a specification given as a regular expression over the set of inputs (or labels associated with transitions) is well understood [27, 129]. Under some assumptions about the regular languages generated by a regular expression, there exist polynomial time (in the size of the transition system) algorithms for deciding the existence of a controller. Such a controller will ensure that the transition system will accept a set of input words (a sublanguage) satisfying the specification. However, for our robotic application, the specification is given over Y, rather than over the set of inputs. This is not a problem if T_S is deterministic, since we can easily transform the problem by labeling all transitions entering a state with the name of that state.

B. For non-terminating robot behaviors (as, for example, in surveillance tasks), *ω-regular expressions* over Y might be a good choice. Such expressions encode ω-regular languages (sets of infinite length strings), which can be accepted by Büchi automaton [129]. Examples include

- *surveillance tasks* ("visit y_1 and y_2 infinitely often", expressed by $(y_2 + y_3 + \cdots + y_{|Y|})^* \cdot (y_1 \cdot (y_1 + y_3 + \ldots + y_{|Y|})^* \cdot y_2 \cdot (y_2 + y_3 + \ldots + y_{|Y|})^*)^\omega$, where the symbol $^\omega$ stands for infinite repetitions).

For designing controllers enforcing such specifications, in [129] it is shown how a controller for T_S can be synthesized (in some cases) from an ω-regular expression over the set of inputs. However, the difficulty of mapping a specification given over the set Y to a specification over the set of inputs still persists, as in the case of regular expressions.

C. *Boolean-based formulas* use the standard logical symbols ¬ (negation), ∧ (conjunction), and ∨ (disjunction). Boolean-based formulas will be used (in Section

6.2) for expressing requirements for terminating robot motions, i.e. each robot will eventually stop in a final state. These expressions will be formulated over symbols from Y, but by using different meanings for an upper-case output (as Y_1) and for its lower-case symbol (y_1) - hence the name "Boolean-based" that we use instead of "Boolean". An upper-case output indicates a requirement along robot trajectories, while a lower-case output includes a requirement for the final robot positions. For example, the formula $\neg Y1 \wedge Y2 \wedge y3$ requires that region O_1 (with output O_2 (output y_2) is visited along the trajectory, and a robot stops in region O_3 (output y_3).

D. *Temporal logics* appeared as frameworks for specifying and verifying the correctness of computer programs, and then they have become used in many other fields due to their resemblance natural language. Formulas of temporal logics are good candidates for robot motion specifiers for three main reasons. First, they are expressive and close to natural language. Second, algorithms for deciding the truth value of formulas (model checkers) are readily available [37, 97]. Third, formulas of temporal logic are usually over the symbols Y that form the outputs of a transition system T_S, rather than over the input set, as in regular and ω-regular expressions.

D.1. *Linear Temporal Logic* (LTL) [10, 14, 37, 61] syntax and semantics, as well as task examples, will be detailed in Section 4.7. For now, we will just mention that, although the expressiveness of LTL formulas is included in that of ω-regular expressions, LTL is more attractive because of its closeness to the natural language and because the specifications are given in terms of states or outputs, rather than over inputs. As will be seen later, the existing LTL model checkers might not help in designing a controller guaranteeing satisfaction of the imposed task. Thus, one of the main challenges will be the development of control strategies from LTL specifications, for both the case of transition systems T_S and T_T and for Petri nets.

D.2. *Computation Tree Logic* (CTL) [37] can also be used as specification language for mobile robot tasks. CTL is also called a branching-time temporal logic, since, in addition to the temporal operators used in LTL, it allows quantification over the paths that are possible from a given state, by using the path operators **A** ("for all runs") and **E** ("for some runs"). However, formulas of CTL are restricted such that each of the temporal operators must be immediately preceded by a path quantifier. LTL and CTL are incomparable, in the sense that there are formulas in CTL that are not expressible in LTL and vice-versa. Typical examples include liveness specifications, which are expressible in LTL but not in CTL, and reset properties, which are expressible in CTL but not in LTL. Relevant to robot motion tasks, a convergence task of the type "eventually always y_i" (requiring that a region O_i is eventually reached and never left after that) can be expressed in LTL, but not in CTL. On the other hand, some specifications involving possible behaviors (existential properties), which can be safety verification of robotic tasks (e.g. "is it always possible to eventually satisfy y_i?"), can be captured in CTL, but not in LTL.

Over LTL, CTL seems to have - at first glance - the advantage that model checking is cheaper, that is, it is performed in polynomial time. However, based on empirical results, there is the belief that for formulas expressible both in CTL and LTL, the model checkers perform similarly [204]. A drawback of CTL is that the translation from natural language to CTL formulas (when possible) is prone to errors [152, 183] (this is due to the fact that one must think of all possible runs at the same time), whereas LTL is more intuitive (because the possible runs are considered one at a time). Moreover, it seems that there are more interesting specifications expressible in LTL and not in CTL than vice-versa [204].

CTL* is a unifying framework for LTL and CTL [37]. However, model checking for CTL* is expensive and its expressivity is too complicated for robotics.

Other directions in temporal logics research focus on continuous time and probabilistic verification. In this book, we will embed the continuous behavior of a system into a discrete world (transition system), where we do not consider the exact moments when transitions can take place (the model being similar to synchronous systems). However, temporal logic frameworks were considered also for asynchronous systems, which are the natural model for continuous time behaviors. While many different models capturing continuous time have been proposed, most of the research in continuous-time model checking focuses on the timed automata model [3]. In probabilistic verification [203], the goal is to tell not only that a certain temporal property (e.g. a failure in a system) is possible but also what the probability of this event is. However, in this book we will not consider the continuous time or the probabilistic verification frameworks.

In our view, a specification language is "good" if it is *natural* (close to human language), *expressive* (allows for the specification of a large class of tasks), and the algorithms for analysis and controller synthesis from its specifications have *low complexity*. Motivated by the above explanations, throughout this work we will use Boolean-based expressions and a subclass of LTL, the so-called LTL_{-X} fragment defined in the next section. Section 5.6 gives a solution for robot motion planning based on LTL_{-X} missions, the solution being based on transition system models. Section 6.3 deals with LTL_{-X} specifications and RMPN models. The usage of Boolean-based formulas in planning a robotic team with an RMPN model is detailed in Section 6.2.

4.7 Linear temporal logic

We begin this section by defining the propositional linear temporal logic known as LTL_{-X} [10, 37].

Definition 4.7 A linear temporal logic LTL_X formula over set $\{y_1, y_2, \ldots, y_{|O|}\}$ is recursively defined as follows:

- every output (or atomic proposition) $y_i \in \{y_1, y_2, \ldots, y_{|O|}\}$ is a formula and
- if ϕ_1 and ϕ_2 are formulas, then $\neg\phi_1$, ϕ_1 or ϕ_2, $\phi_1 \mathcal{U} \phi_2$ are also formulas.

The semantics of LTL_X formulas over Y are interpreted over the set 2^Y, which includes the generated outputs of either a transition system T_S (Definition 4.2), T_T (Definition 4.3), or a Petri net system Q (Definition 4.4). The complete connection between transition systems and LTL_X will appear in Chapter 5 while the connection between Petri net systems and Petri net will be given in Chapter 6.

Definition 4.8 The satisfaction of formula ϕ at position $i \in \mathbb{N} \setminus \{0\}$ of word w, denoted by $w_i \vDash \phi$, is defined recursively as follows:

- $w_i \vDash y$ if $y \in w_i$,
- $w_i \vDash \neg\phi$ if $w_i \nvDash \phi$ (where $\nvDash$ denotes the negation of $\vDash$),
- $w_i \vDash \phi_1$ or ϕ_2 if $w_i \vDash \phi_1$ or $w_i \vDash \phi_2$,
- $w_i \vDash \phi_1 \mathcal{U} \phi_2$ if there exist a $j \geq i$ such that $w_j \vDash \phi_2$ and for all $i \leq k < j$ we have $w_k \vDash \phi_1$.

A word w satisfies an LTL_X formula ϕ, written as $w \vDash \phi$, if $w_1 \vDash \phi$.

The symbols $\neg$ and $\vee$ stand for negation and disjunction. The Boolean constants $\top$ (True) and $\bot$ (False) are defined as $\top = y \vee \neg y$ and $\bot = \neg\top$. The other Boolean connectors $\wedge$ (conjunction), $\Rightarrow$ (implication), and $\Leftrightarrow$ (equivalence), are defined from $\neg$ and $\vee$ in the usual way. The *temporal operator* $\mathcal{U}$ is called the "until" operator. Formula $\phi_1 \mathcal{U} \varphi_2$ intuitively means that (over a word) ϕ_2 will eventually become true and ϕ_1 is true until this happens. Two useful additional temporal operators, "eventually" and "always", can be defined as $\Diamond\phi = \top \mathcal{U} \phi$ and $\Box\phi = \neg\Diamond\neg\phi$, respectively. Formula $\Diamond\phi$ means that ϕ becomes eventually true, whereas $\Box\phi$ indicates that ϕ is true at all positions of w. More expressiveness can be achieved by combining the mentioned operators, as we illustrate in the following paragraph. Table 4.2 shows all Boolean and temporal operators that we use in this book together with the constants.

Example 4.3 *The mentioned syntax makes LTL_X very appealing for specifying motion tasks, and some examples are*

- Reachability tasks: *"reach region y_1 eventually" written as* $\phi_1 = \Diamond y_1$,
- Reachability *and* obstacle avoidance tasks: *"reach y_1 eventually, while always avoiding y_2", written as* $\phi_2 = \Diamond y_1 \wedge \Box\neg y_2$,
- Convergence tasks: *"reach y_1 eventually and stay there for all future times"* - $\phi_3 = \Diamond\Box y_1$,

Table 4.2 Constants, Boolean and temporal operators used to define LTL_{-X} formulas.

Boolean operators	$\neg$	negation	Constants	$\top$	True
	$\vee$	disjunction		$\perp$	False
	$\wedge$	conjunction	Temporal operators	$\mathcal{U}$	until
	$\Rightarrow$	implication		$\Diamond$	eventually
	$\Leftrightarrow$	equivalence		$\Box$	always

- Surveillance tasks: *"visit* y_1 *and then* y_2 *infinitely often"* - $\phi_4 = \Box\Diamond(y_1 \wedge \Diamond y_2)$,
- Choice reachability tasks: *"eventually reach either region* y_1 *or* y_2*"* - $\phi_5 = \Diamond(y_1 or y_2)$.

Moreover, if more robots are available, the attainment of disjoint regions at the same time might be of interest, as in "reach y_1 *and* y_2 *eventually"* ($\phi_6 = \Diamond(y_1 \wedge y_2)$).

LTL [37] is richer than LTL_{-X} in the sense that it allows for an additional temporal operator $\bigcirc$, which is called the "next" operator. Formally, the syntax of LTL is obtained by adding "$\bigcirc\phi_1$" to Definition 4.7 and its semantics is defined by adding "$w_i \vDash \bigcirc\phi$ if $w_{i+1} \vDash \phi$" to Definition 4.8. A careful examination of the LTL and LTL_{-X} semantics shows that the increased expressivity of LTL is manifested only over words with a finite number of successive repetitions of a symbol. This property is also known as closure under stuttering of LTL_{-X} [131]. Consider, for example, the words $w = y_1y_2y_3\ldots$, $w' = y_1y_2y_2y_3\ldots$, and $w'' = y_1y_2y_2y_2\ldots$. Then, in LTL, while all w, w', w'' satisfy formula $\bigcirc y_2$, we can distinguish between w and w' with formula $\bigcirc\bigcirc y_2$, which is true for w' and false for w. On the other hand, w'' satisfies $\bigcirc\Box y_2$, which is false for both w and w'. In LTL_{-X}, we can distinguish between w'' and w or w', because formula $\Diamond\Box y_2$ is true for w'' and false for both w and w'. However, we cannot distinguish between w and w', which would require the $\bigcirc$ operator.

Our choice of LTL_{-X} over LTL is motivated by our definition of the satisfaction of a formula by a continuous trajectory (of a robot) and by our approach to finding runs. Specifically, as it will become clear in Chapter 5, a word corresponding to a continuous trajectory will never have a finite number of successive repetitions of a symbol. An intuitive explanation can also be given by the meaning of the "next" operator in continuous time. Here, one should observe that LTL_{-X} can express requirements like "… from region A enter directly into B (without visiting any other region)", by formula $\ldots A\mathcal{U}B$. For simplicity of notation, we use LTL instead of LTL_{-X} in the rest of this book.

The problem of controlling a transition system such that an LTL formula is satisfied is also called an LTL *planning problem*. In searching for a solution to such a problem, one can try to use available implementations of *model checker* algorithms. Given a transition system T and an LTL formula over its set of observations, a model checker algorithm checks if the formula is satisfied over all possible trajectories of T_S. Such algorithms take as input a non-blocking transition system T_S (also called a Kripke structure) and a formula ϕ, and returns the initial states of T_S from which the formula is satisfied. For the initial states from where the formula is not satisfied, a counterexample is returned. If there are blocking states in T_S, then the "stutter extension" rule [65, 97] can be applied, where self-transitions are artificially added to blocking states (the stutter extension acts like a safety feature). Therefore, in the endeavor of controlling a transition system such that an LTL formula is satisfied, one can model check the transition system (by using off-the-shelf model checking tools, such as SPIN [97] and NuSMV [36]) with the negation of the formula, and then use the returned counterexamples as runs satisfying the initial formula. However, there are several problems that might appear when using this approach. First, if the transition system were non-deterministic, there is in general no hope in exactly following a desired run returned by a model checker. Even if the transition system were deterministic, such a run might not be implementable, because of the stutter extension used in the model checker. Finally, there is no optimality control over the produced counterexamples, and at most the "shortest" ones can be returned by some model checking algorithms. If one wants to impose costs on transitions of T_S, then a model checker cannot be used for obtaining a run having a minimum cost with respect to some predefined optimality criterion (as it will be the case in some problems we will encounter). Thus, we cannot use a model checker to solve the LTL problem we will be facing, and therefore we will have to derive other planning procedures (presented in Chapters 5 and 6, some of them drawing inspiration from model checking).

Definition 4.9 A Büchi automaton is a tuple $B = (S, S_0, \Sigma_B, \delta_B, F)$, where

- S is a finite set of states,
- $S_0 \subseteq S$ is the set of initial states,
- Σ_B is the input alphabet,
- $\delta_B : S \times \Sigma_B \to 2^S$ is a (non-deterministic) transition function,
- $F \subseteq S$ is the set of accepting (final) states.

For $s_i, s_j \in S$, we denote by $\varrho_B(s_i, s_j)$ the set of all inputs of B that enables a transition from s_i to s_j. Moreover, if input $\omega \in \Sigma_B$ enables a transition from s_i to s_j this will be denoted by $s_i \xrightarrow{\omega} Bs_j$.

The semantics of a Büchi automaton is defined over infinite input words. Let $\omega = \omega_1\omega_2\omega_3 \ldots$ be an infinite input word of automaton B, $\omega_i \in \Sigma_B$, $\forall i \in \mathbb{N} \setminus \{0\}$.

We denote by $\mathcal{R}_B(\omega)$ the set of all initialized (i.e. starting from an initial state) runs of B that can be generated by ω:

$$\mathcal{R}_B(\omega) = \{\rho = s_1 s_2 s_3 ... | s_1 \in S_0,\ s_{i+1} \in \delta_B(s_i, \omega_i),\ \forall i \in \mathbb{N} \setminus \{0\}\} \tag{4.13}$$

Definition 4.10 A word ω is accepted by the Büchi automaton B (the word satisfies the automaton) if and only if $\exists \rho \in \mathcal{R}_B(\omega)$ so that $inf(\rho) \cap F \neq \emptyset$, where $inf(\rho)$ denotes the set of states appearing infinitely often in the run ρ.

In [213], it was proved that, for any LTL formula φ over set $Y = \{y_1, y_2, \dots, y_{|O|}\}$, there exists a Büchi automaton B with an input alphabet $\Sigma_B = 2^Y$ accepting *all and only* the infinite strings over 2^Y satisfying formula φ. Since symbols y_i correspond to regions of interest from set O, and the defined set of possible outputs of models in Definitions 4.2, 4.3, and 4.4 is $H \subseteq 2^Y$, we will be able to interpret LTL formulas over model output sequences, as will be detailed in the subsequent chapters.

Moreover, it was proven that if B accepts at least one word, then there exist accepted words that have a *prefix-suffix* structure, consisting of a finite sequence of states (called *prefix*) followed by infinite repetitions of another finite sequence (called *suffix*[2]) [10, 213].

Translation algorithms from LTL formulas to Büchi automata were proposed in [213] and efficient implementations were developed in [73]. The interested reader is referred to [212] for a detailed tutorial on this matter. RMTool uses the algorithm in [73] implemented and available online at http://www.lsv.fr/gastin/ltl2ba/ when it is necessary to obtain the Büchi automaton accepting all strings that satisfy an LTL formula.

Figures 4.8(a) to (f) provide the Büchi automata corresponding to the LTL formulas given in Example 4.3.

In the case when an LTL formula over a set of regions Y expresses the mission for a team of robots, some inputs on the transitions of the Büchi automaton cannot be generated by the robotic system. This is because B has an input alphabet $\Sigma_B = 2^Y$, while the robot model (in the form of either a transition system or RMPN) has the set of possible outputs $H \subseteq 2^Y$, as in Definitions 4.2, 4.3, and 4.4. In order to remove the inputs of B that cannot appear in the team model, we use Algorithm 4.4. Basically, for each transition of B, the set of inputs enabling that transition is updated based on H. Note that some transitions of B may disappear after running Algorithm 4.4.

Example 4.4 *For a better understanding of Algorithm 4.4, considers the LTL formula* $\phi_4 = \Box\Diamond(y_1 \wedge \Diamond y_2)$.

2 Some works refer to "suffix" as being an infinite string resulting from repetitions of a finite one, while we chose to call "suffix" the finite pattern that is infinitely repeated to obtain such an infinite string.

Algorithm 4.4: Update the Büchi automaton B.

Input: Set H - possible observations generated by the robotic model, Büchi automaton B

Output: Trimmed Büchi B.

forall $s_i, s_j \in S$ **do**

  $\varrho_B(s_i, s_j) = \varrho_B(s_i, s_j) \cap H$;

Return updated automaton B;

Let us assume first two robots in the team. The Büchi automaton obtained after running Algorithm 4.4 is the same with the initial one and is given in Figure 4.8(d). The set of possible observations is in this case $H = \{\{y_1\}, \{y_2\}, \{y_1, y_2\}\} = 2^Y$ *and by running Algorithm 4.4, B remains unchanged.*

However, if we assume only one single robot and again disjoint regions of interest, the resulting B after running Algorithm 4.4 is given in Figure 4.9. One robot cannot visit the disjoint regions y_1 *and* y_2 *at the same time, having in this case the set of possible observations* $H = \{\{y_1\}, \{y_2\}, \{\emptyset\}\}$ *(denoted by disjunction* $y_j \vee y_k$*, since* $\emptyset$ *corresponds to no region of interest and hence is not important for transitions of B).* ■

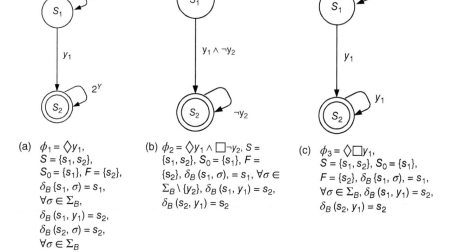

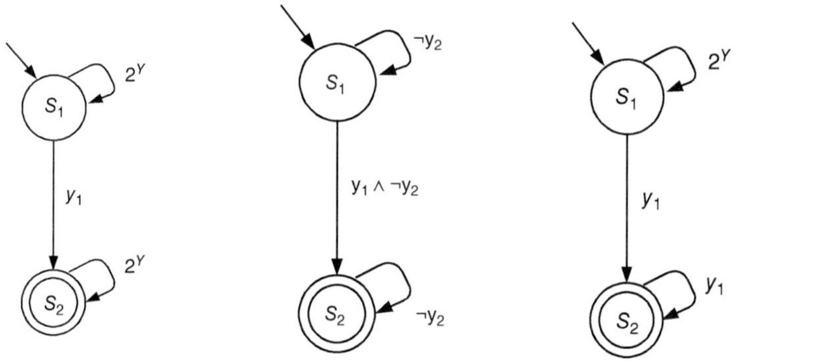

(a) $\phi_1 = \Diamond y_1$, $S = \{s_1, s_2\}$, $S_0 = \{s_1\}$, $F = \{s_2\}$, $\delta_B\{s_1, \sigma) = s_1$, $\forall \sigma \in \Sigma_B$, $\delta_B(s_1, y_1) = s_2$, $\delta_B(s_2, \sigma) = s_2$, $\forall \sigma \in \Sigma_B$

(b) $\phi_2 = \Diamond y_1 \wedge \Box \neg y_2$, $S = \{s_1, s_2\}$, $S_0 = \{s_1\}$, $F = \{s_2\}$, $\delta_B(s_1, \sigma), = s_1$, $\forall \sigma \in \Sigma_B \setminus \{y_2\}$, $\delta_B(s_1, y_1) = s_2$, $\delta_B(s_2, y_1) = s_2$

(c) $\phi_3 = \Diamond \Box y_1$, $S = \{s_1, s_2\}$, $S_0 = \{s_1\}$, $F = \{s_2\}$, $\delta_B\{s_1, \sigma), = s_1$, $\forall \sigma \in \Sigma_B$, $\delta_B(s_1, y_1) = s_2$, $\delta_B(s_2, y_1) = s_2$

Figure 4.8 Examples of Büchi automata corresponding to the LTL formulas given in Example 4.3. The input alphabet is $\Sigma_B = 2^Y = 2^{\{y_1, y_2, \ldots, y_{|O|}\}}$.

Figure 4.8 (Continued)

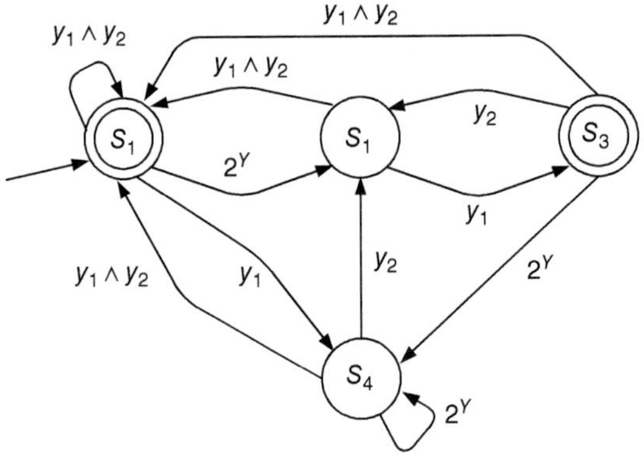

(d) $\phi_4 = \Box\Diamond(y_1 \wedge \Diamond y_2)$, $S = \{s_1, s_2, s_3, s_4\}$, $S_0 = \{s_1\}$, $F = \{s_1, s_3\}$, inputs enabling transitions from δ_B are illustrated on each arrows

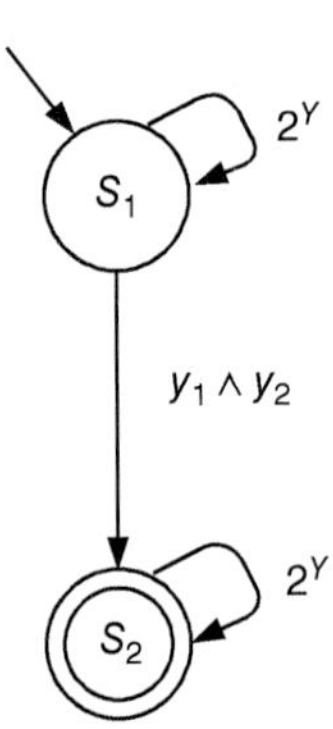

(e) $\phi_5 = \Diamond\ (y_1 \wedge y_2)$, $S = \{s_1, s_2\}$, $S_0 = \{s_1\}$, $F = \{s_2\}$, $\delta_B \{s_1, \sigma), = s_1$, $\forall \sigma \in \Sigma_B$, $\delta_B (s_1, \{y_1, y_2\}) = s_2$, $\delta_B (s_2, \sigma) = s_2$, $\forall \sigma \in \Sigma_B$

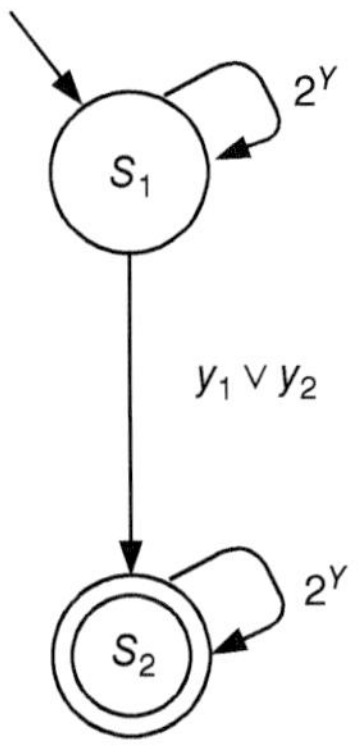

(f) $\phi_6 = \Diamond(y_1 \vee y_2)$, $S = \{s_1, s_2\}$, $S_0 = \{s_1\}$, $F = \{s_2\}$, $\delta_B (s_1, \sigma) = s_1$, $\forall \sigma \in \Sigma_B$, $\delta_B (s_1, \sigma) = s_2$ if $y_1 \in \sigma$ or $y_2 \in \sigma$, $\delta_B (s_2, \sigma) = s_2$, $\forall \sigma \in \Sigma_B$

4.8 Conclusions

Different discrete event systems models that will be used in the following chapters are presented in this chapter. Sections 4.3 and 4.4 define transition systems and Petri net models for modeling a robot or a team of robots. In the case of few robots,

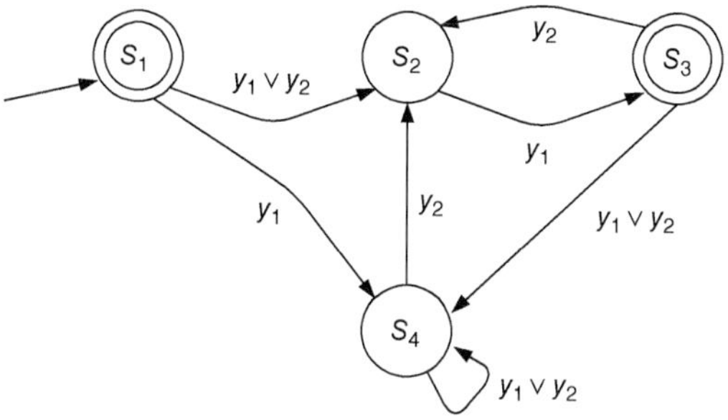

Figure 4.9 Reduced Büchi automaton. Büchi automaton B form of Figure 4.8(d) after running Algorithm 4.4 and assuming one robot in Q and disjoint regions of interest: the set of possible outputs is $H = \{\{y_1\}, \{y_2\}\}$. Note that s_1 does not have a self-loop any more and other transitions disappear.

transition systems are usually preferable. However, if the number of robots in the team is greater than 2 or 3, then Petri net models should be used in order to overcome the state space explosion problem that appears in transition systems due to the synchronous product of more models. Section 4.5 presents a different Petri net model that can be used after the trajectories are computed by a path planning algorithm in order to avoid collisions. This class of Petri nets appeared first in Resource Allocation Systems, while in our case regions are seen as resources. Section 4.6 addressed different formal formalisms for expressing high-level specifications while Section 4.7 focused on LTL and introduced the Büchi automaton that accepts all and only the infinite length words that satisfy an LTL formula.

5

Path Planning by Using Transition System Models

5.1 Introduction

This section presents the classical navigation problem in a rectangular environment $Ev \subset \mathbb{R}^2$ defined as the Cartesian product of $[0, x_{max}] \times [0, y_{max}]$, where $x_{max}, y_{max} \in \mathbb{R}$. The planar environment Ev is cluttered with some convex obstacles. The free space of Ev is partitioned by using a cell decomposition algorithm presented in Chapter 3 and modeled with a transition system $T_S = \langle C, c_0, E, Y, \gamma \rangle$ as in Definition 4.2, with $C = \{c_1, c_2, \ldots, c_{|C|}\}$, the set of cells obtained by the discretization, $Y = \emptyset$, and $\gamma = \emptyset$.

In this environment, we assume that an omnidirectional point robot r is deployed in a known initial position, denoted as $\boldsymbol{s}(r) \in Ev$, and this robot should reach a final point in the environment, denoted as $\boldsymbol{t}(r) \in Ev$.

The following notations will be used:

- $\boldsymbol{centr}(c) \in Ev$ - the centroid of cell corresponding to the node c, $\forall c \in C$;
- $f(c_i, c_j)$ - the common segment of adjacent cells $c_i, c_j \in C$;
- $\boldsymbol{mid}(c_i, c_j) \in Ev$ - the middle point of the segment $f(c_i, c_j)$;
- $f^{-1}(\boldsymbol{w}_i) \subseteq C$ - the set of cells containing point $\boldsymbol{w}_i \in Ev$. Notice that $|f^{-1}(\boldsymbol{w}_i)| = 1$ if $\boldsymbol{w}_i$ is an interior point of a cell, $|f^{-1}(\boldsymbol{w}_i)| = 2$ if $\boldsymbol{w}_i$ belongs to a common segment of adjacent cells, and $|f^{-1}(\boldsymbol{w}_i)| \geq 2$ if it is a vertex of a cell.

Section 5.2 provides the planning approach consisting of two steps: (i) compute the sequence of cells that should be traversed by the robot using a Shortest Path Algorithm and (ii) obtain the intermediate points (waypoints) placed on the common facet of each pair of consecutive cells. Section 5.4 describes an approach in which the path planning and the waypoints optimization is done at the same time using a receding horizon strategy.

Path Planning of Cooperative Mobile Robots Using Discrete Event Models, First Edition.
Cristian Mahulea, Marius Kloetzer, and Ramón González.

5.2 Two-step planning for a single robot and reachability specification

The path will be obtained on the graph given by T_S, and for this, to each edge $(c_i, c_j) \in E$, we assign a *weight* denoted as $\omega(c_i, c_j) \in \mathbb{R}^+$. Different arc weights can be considered, namely

i) $\omega(c_i, c_j) = 1$ - unitary costs;
ii) $\omega(c_i, c_j) = \| \boldsymbol{centr}(c_i) - \boldsymbol{centr}(c_j) \|$ - Euclidean distance between centroids of adjacent cells;
iii) $\omega(c_i, c_j) = \| \boldsymbol{centr}(c_i) - \boldsymbol{mid}(c_i, c_j) \|$ - distance between centroid of cell c_i and middle point of segment shared with the next cell c_j;
iv) $\omega(c_i, c_j) = \left(\sum_{c_h \in N(c_i),\ c_h \neq c_j} \| \boldsymbol{mid}(c_h, c_i) - \boldsymbol{mid}(c_i, c_j) \| / |N(c_i)| - 1 \right)$ - average distance between possible entry points in cell c_i and exit point to cell c_j. The possible entry points are the middle points of segments shared with any neighboring cell c_h different from c_j.

Note that the simplest choice (i), followed by a graph search, yields a path with a minimum number of cells, while choices (ii) to (iv) take into consideration some distances that can be traveled inside cell c_i while moving toward c_j, thus trying to minimize the trajectory length. Clearly, the exact distances to be traveled inside each cell are not known, due to the abstraction of the initial environment to a finite graph, where each whole cell becomes a node. This is why we propose the three weights (ii) to (iv), based on some expected distances inside region c_i. Note that choices (iii) and (iv) produce non-symmetrical graph weights.

The weighted graph can be used to run a search on the graph algorithm, e.g. Dijkstra or A^*, in order to compute a minimum cost path from $f^{-1}(\boldsymbol{s}(r))$ to $f^{-1}(\boldsymbol{t}(r))$. This path gives the cells that should be traversed by the robot in order to reach the destination cell. In order to obtain the trajectory for the robot, one common way is to connect the initial point of the robot with the middle point of the first edge that should be crossed and so on until the last middle point is connected with the destination point.

Example 5.1 *Let us consider the environment in Figure 5.1 containing six obstacles and one robot. By applying the triangular decomposition algorithm described in Section 3.2.3, the environment is divided in to a number of* 38 *cells. This partition is used to define the transition system model of the robot, denoted T_S as in Definition 4.2.*

If weights of types (i), (ii), or (iii) are assigned to the edges of the graph and Dijkstra's algorithm is applied, then the obtained path is the one shown in Figure 5.1(a),

$$path_1 = c_{11}, c_{17}, c_{19}, c_{18}, c_{25}, c_4, c_{28}, c_{30},$$

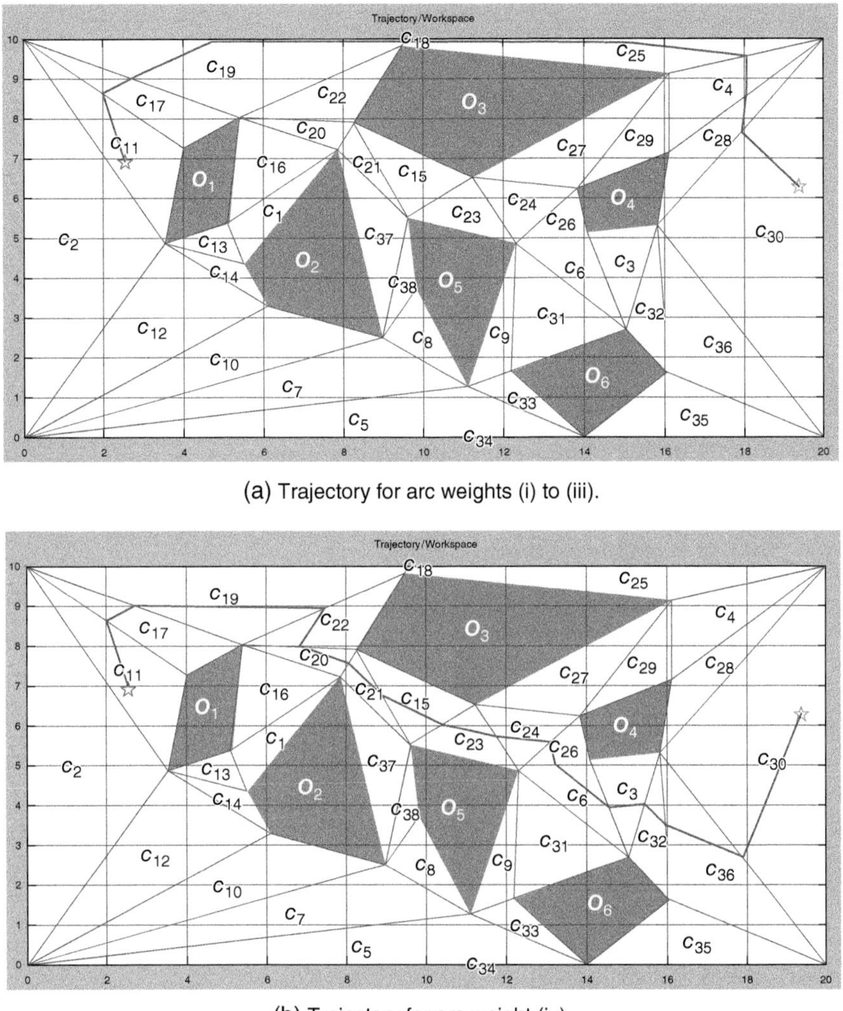

(a) Trajectory for arc weights (i) to (iii).

(b) Trajectory for arc weight (iv).

Figure 5.1 Path planning using a search on a graph (example generated by using RMTool). By assigning different arc weights to the edges of the graph, different trajectories may be obtained.

while the length of the trajectory is equal to 22.12. However, if weights of type (iv) are assigned to the edges, then by applying Dijkstra's algorithm the trajectory shown in Figure 5.1(b),

$$path_2 = c_{11}, c_{17}, c_{19}, c_{22}, c_{20}, c_{21}, c_{15}, c_{23}, c_{24}, c_{26}, c_6, c_3, c_{32}, c_{36}, c_{30},$$

is obtained, which has a length of 25.26.

Once a path in the graph is obtained, we can establish different *waypoints*, i.e. the points on segments (facets) shared by successive cells that yield the piecewise linear robot trajectory. The results presented in [85, 124] used optimization problems to choose the waypoints. Assuming that the initial and final destination are interior points of $c_0 = f^{-1}(\boldsymbol{s}(r))$ and $c_{len} = f^{-1}(\boldsymbol{t}(r))$, let the obtained path be given by the cell sequence

$$path = c_0, c_1, c_2, \dots, , c_{len}$$

and the waypoints to be determined are $\boldsymbol{w}_i \in f(c_{i-1}, c_i)$, $i = 1, \dots, len-1$, where $len + 1$ is the length of the path, i.e. the number of cells of the sequence including initial and final cells. The start point (fixed) is denoted by $\boldsymbol{w}_0 = \boldsymbol{s}(r)$ and the goal point (fixed) is $\boldsymbol{w}_{len} = \boldsymbol{t}(r)$.

Choice (1). Middle points. This is the standard choice [35] and consists of choosing as waypoints the middle points of the common segments crossed by the robot. In particular, the piecewise linear trajectory will be given by the sequence of points

$$traj = \boldsymbol{s}(r), \boldsymbol{mid}(c_0, c_1), \boldsymbol{mid}(c_1, c_2), \dots, \boldsymbol{mid}(c_{len-1}, c_{len}), \boldsymbol{t}(r).$$

Notice that in this case there is no optimization and the waypoints are computed very quickly by using the partition obtained by cell decomposition.

The next choices assume that the piecewise linear trajectory is computed in such a way that the points linking successive cells are obtained in terms of an optimization problem. Let us denote the vertices of the *ith* common line segment (i.e. $f(c_{i-1}, c_i)$, where $\boldsymbol{w}_i \in f(c_{i-1}, c_i)$) by $\boldsymbol{w}'_i$ and $\boldsymbol{w}''_i$. Additionally, a safety margin ϵ is established to avoid the robot passing excessively close to these vertices and possibly to obstacles. The following linear constraints ensure that the waypoint $\boldsymbol{w}_i$ belongs to the line segment $f(c_{i-1}, c_i)$, a segment with vertices $\boldsymbol{w}'_i$ and $\boldsymbol{w}''_i$,

$$\begin{cases} \boldsymbol{w}_i = (1 - \lambda_i) \cdot \boldsymbol{w}'_i + \lambda_i \cdot \boldsymbol{w}''_i, \\ \lambda_i \in \left[\frac{\epsilon}{\|\boldsymbol{w}'_i - \boldsymbol{w}''_i\|}, 1 - \frac{\epsilon}{\|\boldsymbol{w}'_i - \boldsymbol{w}''_i\|} \right], i = 1, \dots, len - 1. \end{cases} \tag{5.1}$$

Choice (2). Manhattan distance. The waypoints are computed by optimizing the sum of L_1 norms of the segments that form the trajectory, or equivalently minimize the Manhattan distance corresponding to the trajectory. This is done by minimizing the cost function

$$J^{L_1 norm} = \sum_{i=0}^{len-1} \| \boldsymbol{w}_{i+1} - \boldsymbol{w}_i \|_1, \tag{5.2}$$

subject to constraints (5.1), $\boldsymbol{w}_0 = \boldsymbol{s}(r)$ and $\boldsymbol{w}_{len} = \boldsymbol{t}(r)$. Optimization problem 5.2 can be transformed to a Linear Programming Problem (LPP) by introducing some new variables

$$\boldsymbol{a}_{(i,i+1)} = |\boldsymbol{w}_{i+1} - \boldsymbol{w}_i|, i = 0, \dots, len - 1,$$

which imply the fulfillment of the two vectorial inequalities:

$$\begin{cases} \boldsymbol{w}_{i+1} - \boldsymbol{w}_i \leq \boldsymbol{a}_{(i,i+1)}, \\ -\boldsymbol{w}_{i+1} + \boldsymbol{w}_i \leq \boldsymbol{a}_{(i,i+1)}. \end{cases} \tag{5.3}$$

By replacing $\boldsymbol{w}_i$ in constraints (5.3) with the convex combination from 5.1, we get the following optimization problem for minimizing $J^{L_1 norm}$:

$$\begin{aligned} & min && \sum_{i=0}^{len-1} \left(\boldsymbol{1}^T \cdot \boldsymbol{a}_{i,i+1}\right) \\ & \text{subject to:} && \\ & && -\lambda_{i+1} \cdot \left(\boldsymbol{w}'_{i+1} - \boldsymbol{w}''_{i+1}\right) + \lambda_i \cdot \left(\boldsymbol{w}'_i - \boldsymbol{w}''_i\right) - \boldsymbol{a}_{i,i+1} \leq \boldsymbol{w}'_i - \boldsymbol{w}'_{i+1}, \\ & && \quad i = 0, \ldots, len-1, \; \lambda_0 = \lambda_{len} = 0 \\ & && \lambda_{i+1} \cdot \left(\boldsymbol{w}'_{i+1} - \boldsymbol{w}'_{i+1}\right) - \lambda_i \cdot \left(\boldsymbol{w}'_i - \boldsymbol{w}''_i\right) - \boldsymbol{a}_{i,i+1} \leq \boldsymbol{w}'_{i+1} - \boldsymbol{w}'_i, \\ & && \quad i = 0, \ldots, len-1, \; \lambda_0 = \lambda_{len} = 0 \\ & && \frac{\epsilon}{\|\boldsymbol{w}'_i - \boldsymbol{w}''_i\|} \leq \lambda_i \leq 1 - \frac{\epsilon}{\|\boldsymbol{w}'_i - \boldsymbol{w}''_i\|}, \; i = 1, \ldots, len-1, \\ & && \boldsymbol{a}_{i,i+1} \geq \boldsymbol{0}, \; i = 0, \ldots, len-1. \end{aligned} \tag{5.4}$$

Optimization (5.4) is a standard LPP [48, 69] with $(len - 1) + 2 \cdot len$ real unknowns ($len - 1$ real λ_i and len bidimensional vectors $\boldsymbol{a}_{i,i+1}$). Note that λ_0 and λ_{len} are two auxiliary constants introduced for the sake of having a unitary notation even for the boundary values of index i. From the optimal solution, we are interested in values λ_i, since these give through (5.1) the unknown points $\boldsymbol{w}_i$, $i = 1, \ldots, len - 1$ along the trajectory. LPPs are considered computationally tractable problems, because the interior-point methods for solving them have a polynomial time complexity [18].

Choice (3). Squared Euclidean distance. This choice considers the optimization of the sum of squared Euclidean distances between successive waypoints, being the cost function defined as

$$J^{sq\_L_2 norm} = \sum_{i=0}^{len-1} \| \boldsymbol{w}_{i+1} - \boldsymbol{w}_i \|^2 = \sum_{i=0}^{len-1} (\boldsymbol{w}_{i+1} - \boldsymbol{w}_i)^T \cdot (\boldsymbol{w}_{i+1} - \boldsymbol{w}_i), \tag{5.5}$$

subject to constraints (5.1), $\boldsymbol{w}_0 = \boldsymbol{s}(r)$ and $\boldsymbol{w}_{len} = \boldsymbol{t}(r)$. By grouping all variables λ_i in the vector $\lambda = [\lambda_1, \ldots, \lambda_{len-1}]^T$, after some mathematical processing the obtained cost function $J^{sq\_L_2 norm}$ can be expressed in the quadratic form

$$J^{sq\_L_2 norm} = \boldsymbol{f} \cdot \lambda + \frac{1}{2} \lambda^T \cdot \boldsymbol{H} \cdot \lambda + R, \tag{5.6}$$

where

$$\boldsymbol{f} \in \mathbb{R}^{len-1}, \; \boldsymbol{f}[i] = 2 \cdot \left(\boldsymbol{w}'_i - \boldsymbol{w}''_i\right)^T \cdot (\boldsymbol{w}'_{i+1} - 2 \cdot \boldsymbol{w}'_i + \boldsymbol{w}'_{i-1}), \; i = 1, \ldots, len - 1, \tag{5.7}$$

$$\begin{aligned}
&\boldsymbol{H} \in \mathbb{R}^{(len-1)\times(len-1)},\\
&\boldsymbol{H}[i,i] = 4\cdot\left(\mathbf{w'}_i - \mathbf{w''}_i\right)^T\cdot\left(\mathbf{w'}_i - \mathbf{w''}_i\right),\ i = 1,\dots,len-1,\\
&\boldsymbol{H}[i,i+1] = -4\cdot\left(\mathbf{w'}_i - \mathbf{w''}_i\right)^T\cdot\left(\mathbf{w'}_{i+1} - \mathbf{w''}_{i+1}\right), i = 1,\dots,len-2,\\
&\boldsymbol{H}[i,j] = 0, i,j = 1,\dots,len-1,\ j \notin \{i,i+1\},
\end{aligned} \tag{5.8}$$

$$R \in \mathbb{R},\ R = \sum_{i=0}^{len-1} \|\ \mathbf{w'}_{i+1} - \mathbf{w'}_i\|^2. \tag{5.9}$$

Term R from Eqs. (5.6) and (5.9) is independent of unknowns λ. Thus, the problem of finding λ_i, $i = 1,\dots,len-1$, with bounds from Eq. (5.1) that minimizes $J^{sq\_L_2norm}$ can be cast in the standard form of a Quadratic Programming (QP) optimization problem [69] as in Eq. (5.10). The cost vector $\boldsymbol{f}$ and matrix $\boldsymbol{H}$ are given by Eqs. (5.7) and (5.8). Since $\boldsymbol{H}$ from Eq. (5.8) is a positive semi-definite matrix, the QP, Eq. (5.10) is a convex optimization and is guaranteed to return a globally optimum solution for $J^{sq\_L_2norm}$, with a polynomial complexity algorithm [126]

$$\begin{aligned}
&\min_{\lambda} \boldsymbol{f}\cdot\lambda + \frac{1}{2}\lambda^T\cdot\boldsymbol{H}\cdot\lambda\\
&\text{subject to:}\\
&\qquad \frac{\epsilon}{\|\ \mathbf{w'}_i - \mathbf{w''}_i\ \|} \le \lambda_i \le 1 - \frac{\epsilon}{\|\ \mathbf{w'}_i - \mathbf{w''}_i\ \|},\\
&\qquad i = 1,\dots,len-1.
\end{aligned} \tag{5.10}$$

Choice (4). Sum of infinity norms. his consists in minimizing the sum of infinity norms of the segments that form the trajectory:

$$\begin{aligned}
J^{L_\infty} &= \sum_{i=0}^{len-1} \|\ \mathbf{w}_{i+1} - \mathbf{w}_i\|_\infty\\
&= \sum_{i=0}^{len-1} \max\left(|\mathbf{w}_{i+1}[1] - \mathbf{w}_i[1]|\ ,\ |\mathbf{w}_{i+1}[2] - \mathbf{w}_i[2]|\right).
\end{aligned} \tag{5.11}$$

Let us denote $\gamma_{(i,i+1)} = \max\left(|\mathbf{w}_{i+1}[1] - \mathbf{w}_i[1]|, |\mathbf{w}_{i+1}[2] - \mathbf{w}_i[2]|\right), i = 0,\dots,len-1$. Thus, similar to Eq. (5.3), we have

$$\begin{cases} \mathbf{w}_{i+1} - \mathbf{w}_i \le \mathbf{1}\cdot\gamma_{(i,i+1)} \\ -\mathbf{w}_{i+1} + \mathbf{w}_i \le \mathbf{1}\cdot\gamma_{(i,i+1)} \end{cases},\ i = 0,\dots,len-1. \tag{5.12}$$

Using expressions (5.1), we obtain the LPP problem from Eq. (5.13) for minimizing $J^{L_\infty norm}$:

$$\min \sum_{i=0}^{len-1} \gamma_{(i,i+1)}$$

subject to:

$$
\begin{aligned}
& -\lambda_{i+1} \cdot \left(\boldsymbol{w}'_{i+1} - \boldsymbol{w}''_{i+1}\right) + \lambda_i \cdot \left(\boldsymbol{w}'_i - \boldsymbol{w}''_i\right) - \boldsymbol{1} \cdot \gamma_{(i,i+1)} \leq \boldsymbol{w}'_i - \boldsymbol{w}'_{i+1}, \\
& i = 0, \dots, len - 1, \ \lambda_0 = \lambda_{len} = 0 \\
& \lambda_{i+1} \cdot \left(\boldsymbol{w}'_{i+1} - \boldsymbol{w}''_{i+1}\right) - \lambda_i \cdot \left(\boldsymbol{w}'_i - \boldsymbol{w}''_i\right) - \boldsymbol{1} \cdot \gamma_{(i,i+1)} \leq \boldsymbol{w}'_{i+1} - \boldsymbol{w}'_i, \\
& i = 0, \dots, len - a, \ \lambda_0 = \lambda_{len} = 0 \\
& \frac{\epsilon}{\parallel \boldsymbol{w}'_i - \boldsymbol{w}''_i \parallel} \leq \lambda_i \leq 1 - \frac{\epsilon}{\parallel \boldsymbol{w}'_i - \boldsymbol{w}''_i \parallel}, \ i = 1, \dots, len - 1 \\
& \gamma_{(i,i+1)} \geq 0, \ i = 0, \dots, len - 1.
\end{aligned}
\tag{5.13}
$$

Choice (5). Euclidean distance. This cost function would minimize the actual length of the trajectory. The problem of finding the intermediate points $\boldsymbol{w}_i$, $i = 1, \dots, len - 1$ that yield the shortest piecewise linear trajectory from $\boldsymbol{w}_0$ to $\boldsymbol{w}_{len}$ included in a given sequence of cells can be formally expressed as follows:

$$
J^{ideal} = \sum_{i=0}^{len-1} \parallel \boldsymbol{w}_{i+1} - \boldsymbol{w}_i \parallel . \tag{5.14}
$$

The nonlinear optimization (5.14) is subject to constraints (5.1). Optimizing an unconstrained sum of Euclidean distances is an NP-hard problem [218]. Unfortunately, to the best of our knowledge, there are no available algorithms that give the globally optimum solution for the constrained optimization (5.14). Thus, one can at most try numerical optimization routines for minimizing Eq. (5.14) with constraints (5.1), but without having a formal guarantee of the solution optimality.

5.3 Quantitative comparison of two-step approaches

For comparing the path planning algorithms with different strategies of computing the waypoints from Section 5.2, we have simulated 1000 random environments with a different number of obstacles varying from 1 to 12; likewise we have different starting and goal positions. All these simulations have been run by using RMTool on a system with Intel G2010@2.8GHz CPU, 4GB RAM, Windows 7 x64, and Matlab 2014a. We consider the next measures (criteria), for each of them, with the value being the average of the 1000 trials, independently for each obstacle number:

- (m1) - Number of traversed cells, depending on graph arc weights (i) to (iv) from Section 5.2;
- (m2) - Time for optimizing trajectory waypoints, i.e. choices (2) to (5) from Section 5.2;

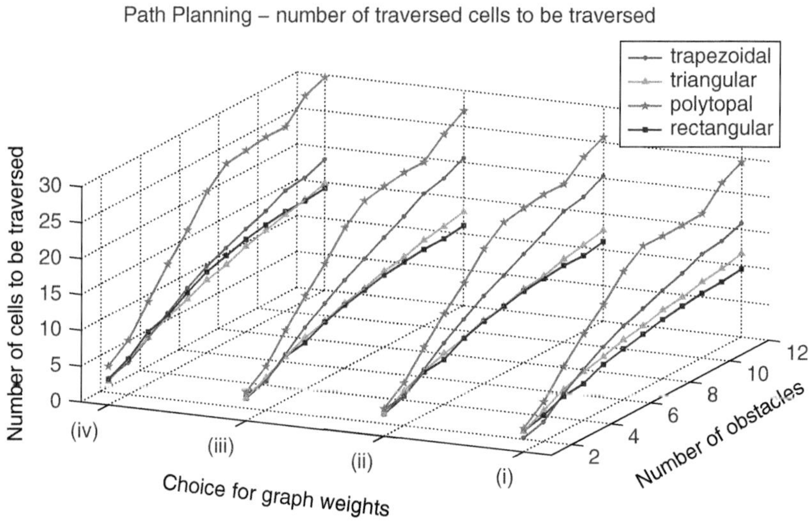

Figure 5.2 Number of cells to be traversed versus number of obstacles and graph weights.

- (m3) - The obtained trajectory length, relative to the shortest possible trajectory (given by the visibility graph). Here we observe how the length is influenced by the planning parameters, i.e., by graph weights (i) to (iv) and by waypoint choices (1) to (5).

Figure 5.2 illustrates measure (m1). For real scenarios, this measure indicates the number of changes in the robot's direction (orientation) or the number of different control laws in traversed cells since for any direction change a new control law should be applied. Especially for rectangular and triangular cells, we observe small increases for all weight choices, relative to the smallest number of cells given by standard choice (i).

Values for measure (m2) are given in Figure 5.3, for weights (i) (time values for other weights are very similar). The non-convex optimization (5) was numerically performed with routine "fmincon" from [196], but the times are not included in Figure 5.3, since they are with more than an order of magnitude higher (up to 0.8 seconds). The other optimizations were performed with standard routines for LPP (waypoints (2) and (4)) and QPP (waypoint (5)) from [196]. Note that all optimization times for choices (2) to (4) are of the order of milliseconds, and thus the computation overhead for including such optimizations is negligible.

Some values for the measure (m3) are illustrated in Figures 5.5, and 5.6. This measure gives very important values and tendencies, showing the increment in percentage of the actual trajectory length versus the minimum possible length

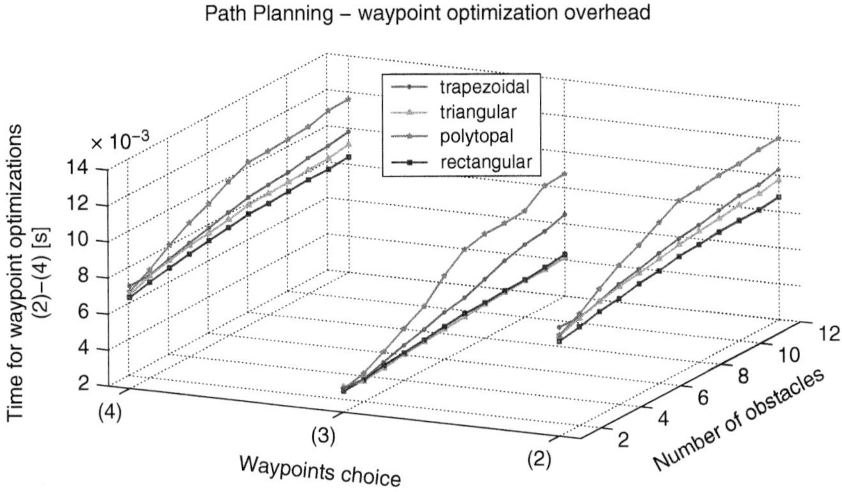

Figure 5.3 Computation time for finding waypoints by optimizations (2) to (4).

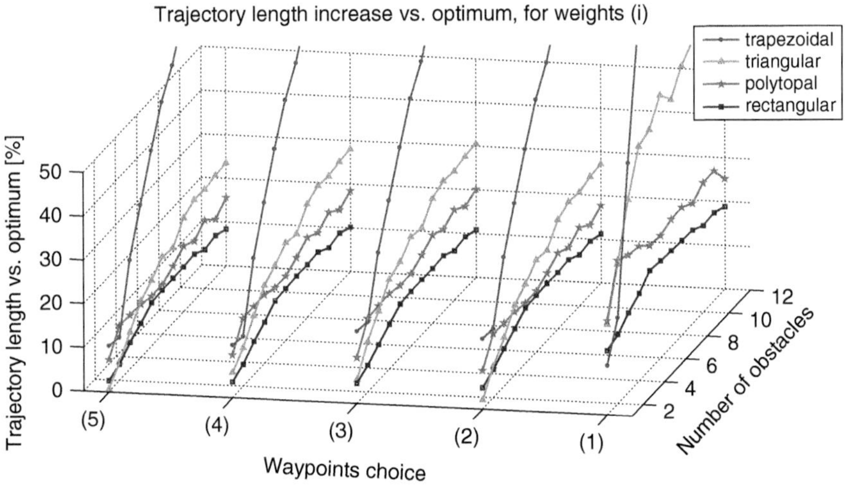

Figure 5.4 Relative increment in length of the obtained trajectory for weights (i). The best result is obtained by the rectangular approach.

given by the visibility graph method. For clarity of representation, the (m3) values higher than 50% are not represented in figures. We mention that in all cases the length increment exceeds 50% only for trapezoidal decompositions, going up to 185% in the case of weights (i) and waypoints (1). Figures 5.4 and 5.5 correspond to weights (i) and (iii), respectively. For weights (ii) (omitted in the figure)

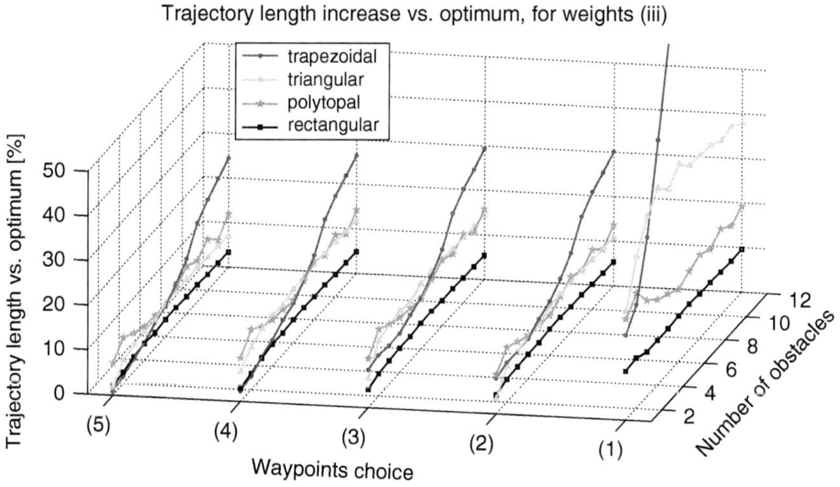

Figure 5.5 Relative increment in length of the obtained trajectory for weights (iii). The best result is obtained by the rectangular approach.

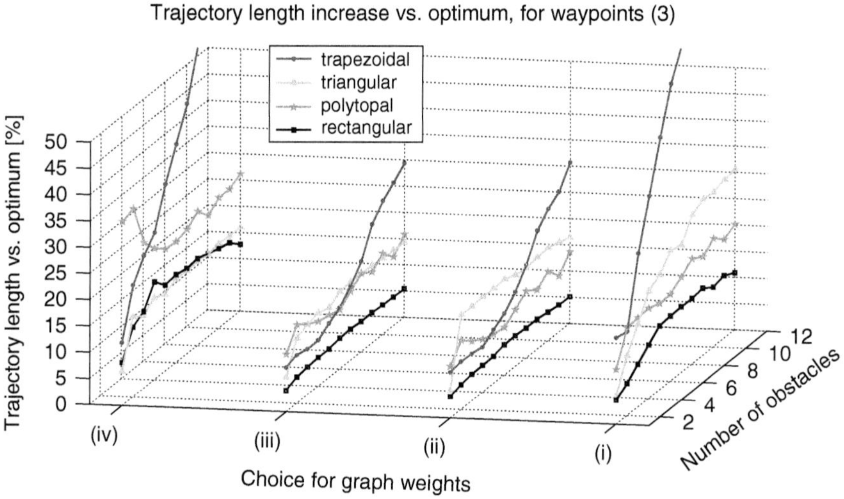

Figure 5.6 Relative increment in length of the obtained trajectory for waypoints choice (3). The best result is obtained by the rectangular approach.

the tendencies are similar to Figure 5.4, but with overall lower values, while for weights (iv) the trapezoidal decomposition yields similar results, but the polytopal one exhibits higher increments in length than the rectangular and triangular ones. Figure 5.6 shows (m3) for optimization (3) versus any weight choice.

We note that, except weights (i) and waypoints (1) and except for trapezoidal decomposition, the relative values of (m3) are below 20%. As noted from other measures, the optimizations (2) to (4) for finding waypoints rely on convex problems already implemented in multiple programming paradigms, while their computation overhead is minimal. Moreover, the number of cells to be traversed is little influenced by weights (ii) to (iv). Except for polytopal cells, the decomposition time and number of cells (also necessary storage memory) are relatively low and show mild increments with the number of obstacles.

According to the results previously described, we consider as good choices for cell decomposition-based planning triangular and rectangular decompositions, with graph weights (ii) and (iii), followed by waypoint optimizations (2), (3), or (4). As observed, the other cases are in general more disadvantageous due to computation time, memory, the length of the trajectory, and the number of piecewise linear segments forming the trajectory.

5.4 Receding horizon approach for a single robot and reachability specification

This section presents another approach for computing a path from the initial point $\boldsymbol{s}(r)$ to the final one $\boldsymbol{t}(r)$ by optimizing the waypoints at the same time with finding traversed cells, rather than the two-step previous method. The solution is based on an optimization problem with a receding horizon strategy.

Let $cost : Ev \times Ev \rightarrow \mathbb{R}$ be the cost function such that for any two points $\boldsymbol{w}_i, \boldsymbol{w}_j \in Ev$, it provides the cost of robot r to move in a straight line from $\boldsymbol{w}_i$ to $\boldsymbol{w}_j$, i.e.

$$cost(\boldsymbol{w}_i, \boldsymbol{w}_j) = \begin{cases} \| \boldsymbol{w}_i - \boldsymbol{w}_j \|, & \text{if the straight line between } \boldsymbol{w}_i, \boldsymbol{w}_j \text{ does not} \\ & \text{intersetct with any obstacle,} \\ \infty, & \text{otherwise.} \end{cases} \tag{5.15}$$

As described in Section 5.2, if the waypoint $\boldsymbol{w}_i$ is located in a variable position within the common edge between two consecutive cells, then the optimization problem becomes nonlinear (choice (5) in Section 5.2). In order to overcome this problem, one possibility used in this section is to consider only a finite number of points on each common edge as possible candidates for the waypoints.

As in the methods presented in Section 5.2, a safety margin ϵ is imposed on each common edge of adjacent cells (near the vertices) in order to avoid small distances between the robot and obstacles. The robot is prohibited to cross the edge through any point on these safety segments. The remaining part of the edge is discretized in a number of S segments. Let $edge(c_i, c_j)$ be the equally spaced finite set of points placed on the common edge of the adjacent cells c_i and c_j, with $edge(c_i, c_j) = \emptyset$ if c_i and c_j are not adjacent or $c_i = c_j$. Let $Edges = \bigcup_{i,j=1}^{|C|} edge(c_i, c_j)$ be the set of points placed on all edges. Finally, for a point $\boldsymbol{w}_k \in Ev$, let $edge^{-1}(\boldsymbol{w}_k) = \{c_i, c_j\}$ be the set of the two cells such that $\boldsymbol{w}_k \in edge(c_i, c_j)$; $edge^{-1}(\boldsymbol{w}_k) = \{c_h\}$ if $\boldsymbol{w}_k$ is an interior point of cell c_h; and $edge^{-1}(\boldsymbol{w}_k) = \emptyset$ for points outside the free space of Ev.

The waypoints are computed by solving iteratively the following optimization problem:

$$\begin{aligned} \min J = & \sum_{i=0}^{N-1} \left(cost(\boldsymbol{w}_i, \boldsymbol{w}_{i+1})\right) + cost_t(\boldsymbol{w}_N, \boldsymbol{t}(r)) \\ s.t. \quad & \boldsymbol{w}_i \in Edges \cup \{\boldsymbol{s}(r), \boldsymbol{t}(r)\}, \quad \forall i = 0, \dots, N \\ & edge^{-1}(\boldsymbol{w}_i) \cap edge^{-1}(\boldsymbol{w}_{i+1}) \neq \emptyset, \forall i = 0, \dots, N-1, \end{aligned} \tag{5.16}$$

where N is the timing horizon while $cost_t(\boldsymbol{w}_N, \boldsymbol{t}(r))$ is the cost from the Nth predicted waypoint to the final point of the robot, $\boldsymbol{t}(r)$.

The first term in the cost function in Eq. (5.16) represents the cost to go from the start point $\boldsymbol{w}_0$ to $\boldsymbol{w}_N$ (first N waypoints), hence considering only a number of N intermediate waypoints. From $\boldsymbol{w}_i$ to $\boldsymbol{w}_{i+1}$ we assume that the robot is going in a straight line and $\boldsymbol{w}_i$ and $\boldsymbol{w}_{i+1}$ are chosen from a finite set of candidate points and belong to the same cell (according to second constraint). The second term instead represents a terminal cost from $\boldsymbol{w}_N$ to $\boldsymbol{t}(r)$. Since several cells could be crossed from $\boldsymbol{w}_N$ to $\boldsymbol{t}(r)$ and possibly some obstacles avoided, it may be difficult to compute the optimal cost in a parametric way to be introduced in the cost function. For example, the minimal trajectory length from $\boldsymbol{w}_N$ to $\boldsymbol{t}(r)$ is not in general the straight line between both points when obstacles are presented in the environment.

Starting from $\boldsymbol{s}(r)$, the optimization problem (5.16) is solved initially with $\boldsymbol{w}_0 = \boldsymbol{s}(r)$. From the solution that is obtained, only $\boldsymbol{w}_1$ is kept and hence the intermediate trajectory will be $\boldsymbol{s}(r), \boldsymbol{w}_1$. Then, the optimization problem (5.16) is solved again with $\boldsymbol{w}_0 = \boldsymbol{w}_1$ and the procedure is iterated until $\boldsymbol{w}_1 = \boldsymbol{t}(r)$ in the solution. However, as in standard MPC, convergence problems may appear and the destination could never be reached. This is usually the case if N is too small and the terminal cost is not correctly chosen. On the other hand, if N is big, the optimization problem converges but the computational complexity of the solution is also very big. Indeed, if N is big, there is an explosion of combinations in terms of possible paths of length N.

Example 5.2 *Let us consider the environment in Figure 5.7 and assume $N = 2$ and the initial position of the robot is as it is shown in the figure. Furthermore, let $cost_t(\boldsymbol{w}_N, \boldsymbol{t}(r)) = \| \boldsymbol{w}_N - \boldsymbol{t}(r) \|$, i.e. Euclidean distance between both points.*

The optimal solution of the problem (5.16) selects the next intermediate point on the edge placed in the direction of the arrow. This is because the terminal cost $cost_t$ to the target region is the Euclidean distance and the cost from the cell placed at the right is

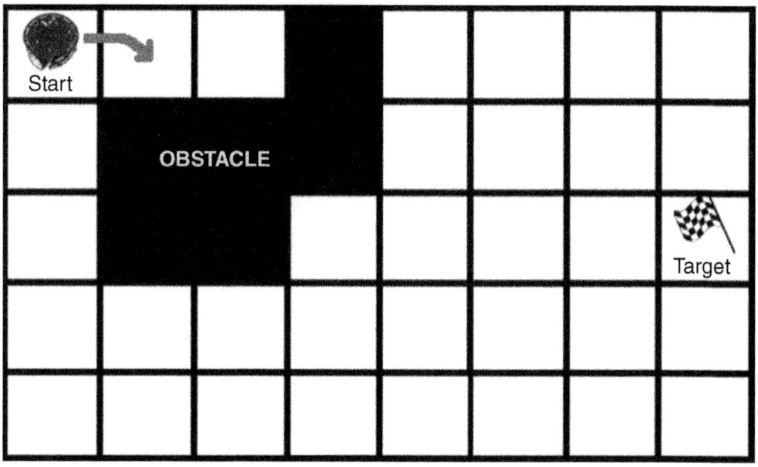

Figure 5.7 Environment used in Example 5.2. If N is too small and the terminal cost is not correctly chosen, the target point is never reached.

much smaller than the cost from the cell placed bottom from the starting point of the robot. If the procedure is iterated, the robot will go right until it detects the obstacle; then it will go back and so on. The robot will never reach the final destination. The problem may be solved by increasing N or by changing the terminal cost in the cost function. ■

By applying the Dijkstra's Shortest Path Algorithm [53] on the graph T_S modeling the robot movement capabilities by using as start node $edge^{-1}(\boldsymbol{t}(r))$, i.e., the cell corresponding to the final destination of the robot, one can compute the minimum distance from node $edge^{-1}(\boldsymbol{t}(r))$ to all other nodes on the graph. The weights of the edges in the graph T_S could be any choice (i) to (iv) from Section 5.2. In RMTool, weights of type (ii) are considered, i.e., Euclidean distance between the centroids of the cells. Since the graph is undirected, the minimum distances are equal also to the distance from a given node to the destination one. For this, only one execution of the Dijkstra's Shortest Path Algorithm is required. Let $dist(c_i)$ be the minimum distance between cell c_i and $edge^{-1}(\boldsymbol{t}(r))$ obtained by applying the Search algorithm. Let us define the terminal cost in (5.16) as

$$cost_t(\boldsymbol{w}_N, \boldsymbol{t}(r)) = M \cdot \min_{c_i \in edge^{-1}(\boldsymbol{w}_N)} \left(dist(c_i) + ||\boldsymbol{w}_N - \boldsymbol{centr}(c_i)|| \right), \tag{5.17}$$

where C is a big number. In the following simulations, $M = 1000$ is used for an environment of 20×10. Notice that by this choice of the terminal cost, a big weight C is given to the minimal distance from the cell corresponding to $\boldsymbol{w}_N$ to the destination plus the distance to the centroid of the cell corresponding to $\boldsymbol{w}_N$. In this way, the problem that we have in Example 5.2 will not appear since this choice

Algorithm 5.1: Path planning optimizing the waypoints

Input: Partition C, N, $edge(c_i, c_j), \forall c_i, c_j \in C$, $\mathbf{s}(r)$ and $\mathbf{t}(r)$

Output: *trajectory*

1 Compute the minimum cost $dest(c_i)$ for each cell $c_i \in C$ to reach the destination cell $edge^{-1}(\mathbf{t}(r))$, on graph T_S with weights of type (ii) from Section 5.2.

2 Let $\mathbf{w}_0 = \mathbf{s}(r)$;

3 Let $trajectory = \mathbf{w}_0$;

4 **repeat**

5   Let $reach \subseteq C$ be the set of cells reachable in N steps from all nodes $edge^{-1}(\mathbf{w}_0)$;

6     $Edges = \emptyset$;

7   **for** $c_i \in reach$ **do**

8   **for** $c_j \in reach$ **do**

9             $Edges = Edges \cup edge(c_i, c_j)$;

10   Let $Edges = Edges \cup \{\mathbf{w}_0\}$;

11   Create a graph G', defining for each $w \in Edges$ a node in G' and between nodes $w_i, w_j \in Edges$ add an arc if $edge^{-1}(w_i) \cap edge^{-1}(w_j) \neq \emptyset$ and assign to this edge the cost given by Eq. (5.15).

12   Add to G' a node n' corresponding to the destination point. Connect all other nodes of G' with n' and assign them the cost given by Eq. (5.17).

13   Compute on G' the minimal path $run = \mathbf{w}_0\mathbf{w}_1\mathbf{w}_2 \ldots n'$;

14     $trajectory = trajectory, \mathbf{w}_1$;

15     $\mathbf{w}_0 = \mathbf{w}_1$;

16 **until** $\mathbf{w}_0 = \mathbf{t}(r)$;

of the terminal cost is a very good estimation of the optimal cost from $\mathbf{w}_N$ to $\mathbf{t}(r)$. Since the terminal cost is computed on a graph with weights corresponding to the distance between the centroids of cells, if the destination point is not a centroid cell then the solution may not be optimal. However, the algorithm converges.

Optimization problem (5.16) can be solved by iteratively running another Shortest Path Algorithm on a graph created based on the possible intermediate waypoints. This algorithm has, in general, lower computational complexity than solving a Mixed Integer Linear Programming problem. The steps of the algorithm are synthesized in Algorithm 5.1.

Step 1 of Algorithm 5.1 computes the minimum distance between all cells of C and the cell containing the final destination point. As mentioned before, this is necessary for only one execution of a Shortest Path algorithm. Step 2 initializes $\mathbf{w}_0$ with the initial point of the robot while Step 3 initializes *trajectory* with $\mathbf{w}_0$. Step 4 is the starting point of a loop that finishes when the final point $\boldsymbol{t}(r)$ is reached. The steps of loops are the following. In Step 5, all nodes reachable in a number of N steps from the nodes corresponding to $\boldsymbol{w}_0$ are computed. Notice that if $\boldsymbol{w}_0$ is on an edge, then *reach* will contain all nodes reachable in N steps from both cells that share the edge. Steps 6 to 12 create in variable *Edges* the set of all possible waypoints. Recall that $edge(c_i, c_j)$ is a set of S equally spaced points on the the common edge of cells c_i and c_j (if they are adjacent), otherwise $edge(c_i, c_j) = \emptyset$. In Step 13, a new graph is computed based on the elements in *Edges*. For each element in *Edges* a node in the graph is added and the arcs are added only if the points belong to the same cell of C. Furthermore, the cost assigned to a new arc is given by the Euclidean distance between the corresponding points. Step 14 adds a node n' corresponding to the final destination of the robot and this state is connected to all the others with edges and cost given by the distance computed in Step 1. On the new graph, a shortest path algorithm is applied to go from $\boldsymbol{w}_0$ to the last node added. The trajectory is updated with the first point of the optimal solution, i.e. $\boldsymbol{w}_1$, and the algorithm is iterated until the final destination is reached.

5.5 Simulations and analysis

This section discusses some simulations done by using the RMTool with the final goal to understand the behavior of Algorithm 5.1 when varying some of its conditions, in particular:

- The variation of the cost of the trajectory and the time complexity with respect to the variation of N and S parameters;
- The difference, in terms of cost, by changing the cell decomposition shape used to discretize the environment. In particular, we used four cell decomposition algorithms, namely triangular, rectangular, trapezoidal, and polytopal.

For the first objective, assume the scenario in Figure 5.8 on which three obstacles, one robot, and one target point are placed. The safety margin on the common edge near the vertices is fixed to $\epsilon = 0.5$.

The simulation results are shown in Table 5.1 and it can be observed that with low N values, the S parameter induces only low variations on the time complexity. This is due to the fact that if N (the number of lookahead steps the robot can make) is reduced, the number of paths that can be generated from its current position will not be high, so it is possible to increase the fitting within the common sides of two different cells. Likewise, it can be seen that, as the N increases, the final cost is reduced. Of course, the table also shows how, with the increment of the parameters N and S, the time complexity will not linearly increase, so it is necessary to run several trials in order to identify the best trade-off between these two variables. In

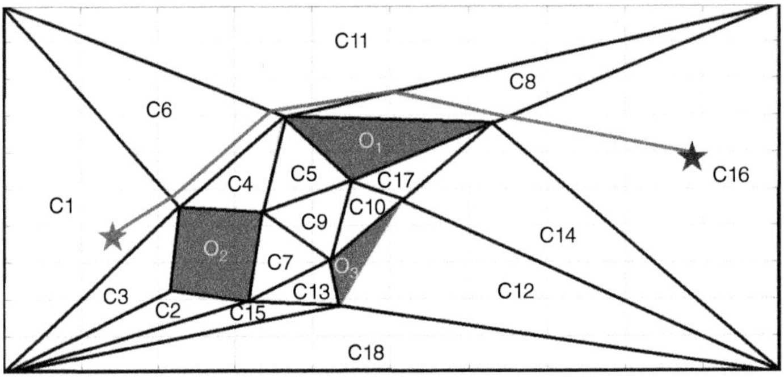

Figure 5.8 Environment of size 20 × 10. Simulation with a receding horizon strategy ($N = 2$ and $S = 5$). Environment containing three obstacles, discretized with the triangular cell decomposition. The start robot position is the red left star while the target position is the blue right star. The red trajectory is obtained with $N = 2$ and $S = 5$ and has a cost equal to 17.54.

this regard, it is also possible to note that, for big N and S values, it is not necessary to further increase these parameters, since the final result will not be significantly affected in terms of cost. Figure 5.8 shows the trajectory obtained for $N = 2$ while Figure 5.9 shows the obtained trajectory for $N = 5$.

It is clear that, with the increment of N, there is a significant reduction of the cost but, simultaneously, there is an increment of the computational complexity. In this regard, it is important to say that this is true when *the target point is placed on the centroid of the cell.* If not, a possible not optimal path could "appear" as advantageous, since the point positioned on the common side between the final cell and an adjacent one could have a smaller distance from the end point to the point belonging to the optimal path.

In support of this statement, we show in Table 5.2 how the final cost changes with the variation of N and S, in case the end point is positioned as in Figure 5.10. In this case, while increasing N, the cost of trajectory is slightly bigger.

Clearly, for other types of cell decomposition methods, these parameters can have a different impact, aspects like the shape of each individual cell can make the number of cells surrounding a single cell variable (increasing or decreasing it), thus also affecting the right value which could assume N and S in order to produce a good solution in terms of cost. In order to verify this last consideration, we took as reference the same scenario as in Figure 5.10 and we changed the cell decomposition method, trying the ones mentioned at the beginning of this section. To make a comparison, we fixed the value of S to 3, which is a good average value, and we observed the differences between the methods with different values of N.

Table 5.3 shows the main differences that appear by using different decomposition approaches. Firstly, we can see that the algorithm gives better performances

Table 5.1 Cost and Time Complexity by varying N and S parameters.

S	N	Cost (cm)	Time Complexity (s)
2	1	23.46	0.02
2	2	19.11	0.03
2	3	19.11	0.06
2	5	16.61	0.90
2	10	16.61	1.04
2	20	16.61	1.21
3	1	20.96	0.02
3	2	17.01	0.07
3	3	17.01	0.11
3	5	15.32	1.54
3	10	15.32	1.77
3	20	15.32	2.35
4	1	20.88	0.04
4	2	17.34	0.09
4	3	17.34	0.20
4	5	15.53	3.48
4	10	15.53	3.84
4	20	15.32	5.85
5	1	18.79	0.05
5	2	17.54	0.14
5	3	17.54	0.29
5	5	15.80	6.02
5	10	15.80	6.43
5	20	15.80	8.32

with the triangular cell decomposition. Furthermore, we can note how, using the rectangular cell decomposition, the time complexity explodes with the increment of N, due to the number of cells that this decomposition technique produces.

Finally, another interesting aspect that is important to analyze is the one related to the differences between the different approaches presented in Section 5.2 and the one using the receding horizon presented in Section 5.4. Fixing $N = 10$, we will check the differences in terms of performances between these approaches. Table 5.4 shows that the proposed algorithm has the best performances, in terms of cost, when compared to the other solutions. Increasing the number of regions

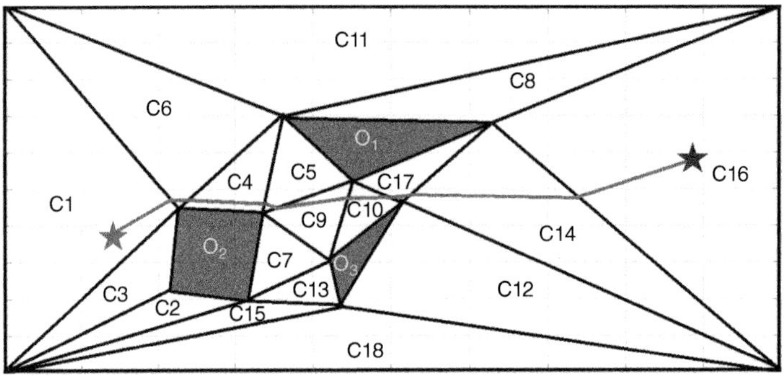

Figure 5.9 Simulation with a receding horizon strategy ($N = 5$ and $S = 5$). Resulting trajectory with $N = 5$ and $S = 5$, with the cost equal to 15.80.

Table 5.2 Cost and Time Complexity by varying N and S parameters when the final point is not placed in the centroid of the cell.

S	N	Cost (cm)	Time Complexity (s)
2	2	16.76	0.07
2	10	19.94	0.89
2	20	19.94	1.54
3	2	17.31	0.10
3	10	18.02	1.56
3	20	18.02	3.04
4	2	16.90	0.20
4	10	19.81	2.38
4	20	19.81	4.20
5	2	16.86	0.26
5	10	18.65	3.41
5	20	18.65	6.19

within the scenario, the time complexity is greater then the other methods, and for that reason the tuning of the N and S parameters is a critical aspect to take into consideration.

5.6 Path planning with an $LTL_{\_X}$ specification

In this section, the robotic mission is specified as a Linear Temporal Logic ($LTL_{\_X}$) formula, often used for allowing expressive control objectives [14, 66, 111, 141, 214]. Tasks given in this logic usually imply visits to some regions of interest from

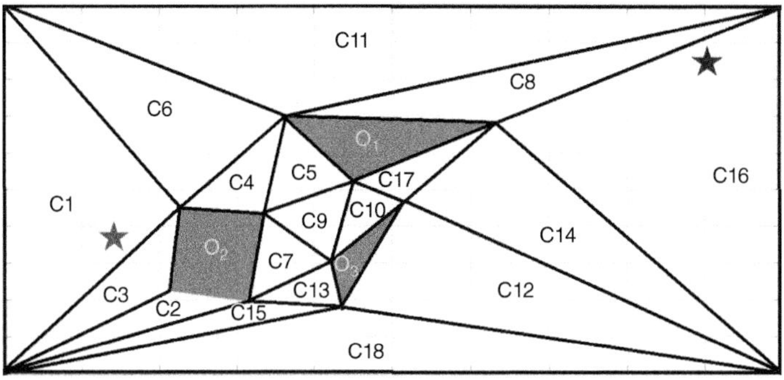

Figure 5.10 Scenario with the final point (blue star on the right) not placed on the centroid of the cell.

Table 5.3 Cost and Time Complexity by varying N for each different cell decomposition method, fixing S equal to 3.

Decomp.	N. Regions	N	Cost (cm)	Time Complexity (s)
Triangular	18	2	17.01	0.07
Triangular	18	10	15.32	1.77
Triangular	18	20	15.32	2.35
Rectangular	113	2	17.26	9.48
Rectangular	113	10	16.55	462.09
Rectangular	113	20	16.55	473.56
Trapezoidal	14	2	20.52	0.19
Trapezoidal	14	10	17.41	1.01
Trapezoidal	14	20	17.41	1.03
Polytopal	40	2	16.39	0.84
Polytopal	40	10	16.23	14.46
Polytopal	40	20	16.23	14.53

the environment in a specific or arbitrary order, avoidance of other regions, and various logical and temporal connectives among regions.

Recall that Section 4.7 provides an introduction to $LTL_{\_X}$ formulas and the corresponding Büchi automata that accept as input infinite words that satisfy the formulas. As explained, we further use notation LTL instead of $LTL_{\_X}$, since the omitted "next" operator in meaningless for continuous trajectories, such as those generated by mobile robots.

We detail first the case of planning a single robot based on an LTL formula, and then the solution is extended to a team of identical robots.

Table 5.4 Comparison of the performances by using different Intermediate Trajectory Points approaches for each cell decomposition method.

Decomp.	Int.Traj.Point	Cost (cm)	Time Complexity (s)
Triangular	Mid.Point (choice (1))	20.03	0.01
Triangular	Norm.1 (choice (2))	20.19	1.17
Triangular	Sq. Norm.2 (choice (3))	16.57	0.28
Triangular	Norm.Inf. (choice (4))	16.92	0.05
Triangular	MPC (receding horizon)	15.32	1.77
Rectangular	Mid.Point (choice (1))	17.60	0.01
Rectangular	Norm.1 (choice (2))	17.72	0.03
Rectangular	Sq. Norm.2 (choice (3))	17.20	0.02
Rectangular	Norm.Inf. (choice (4))	18.37	0.02
Rectangular	MPC (receding horizon)	16.55	462.09
Trapezoidal	Mid.Point (choice (1))	23.21	0.01
Trapezoidal	Norm.1 (choice (2))	22.16	0.02
Trapezoidal	Sq. Norm.2 (choice (3))	20.23	0.02
Trapezoidal	Norm.Inf. (choice (4))	20.46	0.02
Trapezoidal	MPC (receding horizon)	17.41	1.01
Polytopal	Mid.Point (choice (1))	19.18	0.01
Polytopal	Norm.1 (choice (2))	19.32	0.07
Polytopal	Sq. Norm.2 (choice (3))	17.37	0.01
Polytopal	Norm.Inf. (choice (4))	18.29	0.03
Polytopal	MPC (receding horizon)	16.23	14.46

Single-robot case. The automatic planning strategy begins by converting the imposed LTL specification into a Büchi automaton $B = \langle S, S_0, \Sigma_B, \delta_B, F\rangle$ as in Definition 4.9, by using the software introduced in [73]. The input alphabet Σ_B of B is the output alphabet of the transition system modeling the movement capabilities of the robot (transition system $T_S = \langle C, c_0, E, Y, \gamma\rangle$ in Definition 4.2). Therefore, $\Sigma_B = Y$. A special type of product automaton between T_S and B is computed and will be used to search an optimal path (run).

Definition 5.1 The product automaton $A = T_S \times B$ is constructed as follows: $A = \langle S_A, S_{A0}, \delta_A, F_A\rangle$, where:

- $S_A = C \times S$ is the set of states;
- $S_{A0} = c_0 \times S_0$ is the set of initial states;

- $\delta_A \subseteq S_A \times S_A$ is the transition relation, defined by: $\big((c,s),(c',s')\big) \in \delta_A$ if and only if $(c,c') \in E$ and $s' \in \delta_B(s,\gamma(c))$, where $c, c' \in C$, $s, s' \in S$;
- $F_A = C \times F$ is the set of final (accepting) states.

The product automaton A, sometimes referred to as a synchronous product [97], represents a connection robot capabilities (modeled by T_S) and formula satisfaction (witnessed by B) [14]. Thus, a transition in A represents a matching condition between a transition of T_S and a transition of B caused by the current observation of T_S, as shown by relation δ_A [1].

The acceptance condition of A is formulated similar to the one of B, in the sense that an infinite run is accepted by A if and only if it visits infinitely often the set of final states F_A. As mentioned for B in Section 4.7, if A has at least one accepted run, then it has an accepted run in a *prefix-suffix* form, i.e. formed by a finite sequence of states (prefix) followed by infinite repetitions of another finite sequence (suffix).

For finding an accepted run of A, denoted $path_A$, we use Algorithm 5.2 (lines 1-16) [109]. The prefix and suffix composing $path_A$ are determined by using standard graph search algorithms (e.g. Dijkstra [42, 53]), denoted by the routine *find_path* in Algorithm 5.2. The *prefix* is the part of the path that starts from an initial state and reaches a state s_{F_A} from set F_A, while the corresponding *suffix* repeatedly visits s_{F_A}. Note that for finding the suffix (starting and returning to s_{F_A}) a graph search cannot be called with an identical start and destination node, because this would always output a path, even if no self-loop is in that node. Therefore, we use the idea from lines 9-16, i.e. searching paths from s_{F_A} to all states s_{neigh} that can transit in s_{F_A} and choosing the minimum cost path. For calling routine *find_path*, one can use different weights to transitions from δ_A (e.g. inherited from transition weights in T_S), or simply use unitary weights. These weights yield the total cost of a path of A, used on lines 14-15 via routine *total_cost*. Algorithm 5.2 chooses an accepted path of A with a minimum cost of prefix and suffix. For simplicity, we can use unitary weights and thus obtain an accepted run of A with a minimum number of states in the prefix followed by the suffix. Obviously, the cost function on lines 14-15 can be altered, e.g. by assuming several repetitions of the suffix. Once an accepted path of A is obtained in Algorithm 5.2, it is projected to states of T_S, i.e. only the states from C are kept. Thus, $path_{T_S}$ is obtained, with the guarantee that its generated output sequence satisfies the LTL formula. Clearly, $path_{T_S}$ has the same prefix-suffix structure as the accepted run of A.

For moving the mobile robot, the sequence of cells $path_{T_S}$ returned by Algorithm 5.2 is converted to a trajectory, e.g. by choosing waypoints as in the

---

1 An alternative definition of δ_A can use the next observation of T_S instead of the current one, while adding an initial dummy state to T_S for accounting output in c_0 [65, 111].

Algorithm 5.2: Find a path of T_S satisfying an LTL formula

Input: T_S - robot model, B - Büchi automaton corresponding to LTL formula
Output: $path_{T_S}$ - run in T_S whose output sequence satisfies the LTL specification

Construct $A = T_S \times B$ as in Definition 5.1
$path_A = \emptyset$
$min\_cost\_run = \infty$
for $s_{A0} \in S_{A0}$ **do**
    for $s_{F_A} \in F_A$ **do**
        $prefix = find\_path(source: s_{A0}, destination: s_{F_A})$
        if $prefix \neq \emptyset$ **then**
            Remove last state from $prefix$ (s_{F_A})
            for $s_{neigh} \in S_A$ $s.t.$ $\left(s_{neigh}, s_{F_A}\right) \in \delta_A$ **do**
                if $s_{neigh} == s_{F_A}$ **then**
                    $suffix = s_{F_A}$
                else
                    $suffix = find\_path(source: s_{F_A}, destination: s_{neigh})$
                if $suffix \neq \emptyset$ and $total\_cost(prefix, suffix) < min\_cost\_run$ **then**
                    $min\_cost\_run = total\_cost(prefix, suffix)$
                    $path_A = prefix, suffix, suffix, \ldots$
if $path_A \neq \emptyset$ **then**
    Project $path_A$ to sequence of states from C, obtaining $path_{T_S}$
else
    No path of T_S can satisfy the LTL formula

second step of approach from Section 5.2. Then the robot should follow the prefix and then infinitely repeat the suffix (of course, in real scenarios the number of suffix repetitions would be limited). If the suffix consists in one state of T_S, the robot stops in the corresponding cell.

Example 5.3 *Consider the environment from Figure 4.1, with only the first robot, initially deployed in cell c_1. Recall that the transition model T_S for the evolution of this robot in the given environment was presented in Figure 4.3. Assume the LTL specification $\phi_3 = \Diamond\Box y_1$, whose Büchi automaton B is depicted in Figure 4.8(c).*

For this example, the product automaton A constructed as in Definition 5.1 is illustrated in Figure 5.11. Automaton A has 12 states, while the initial state is (c_1, s_1). There are six accepting states, double encircled in Figure 5.11,

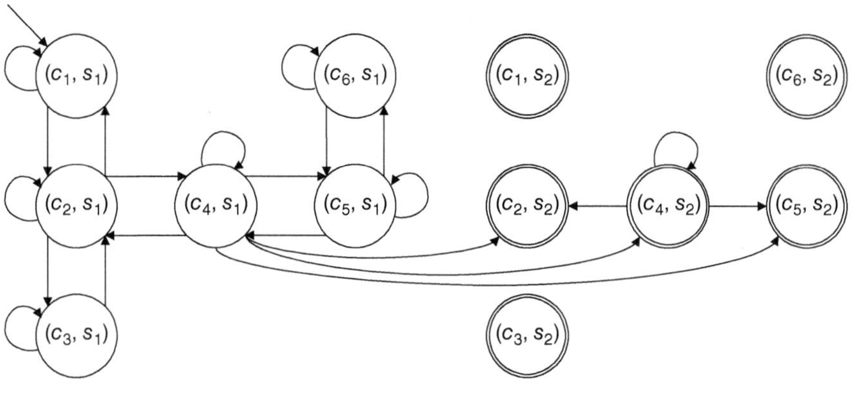

Figure 5.11 Example of a product automaton between the transition system T_S from Figure 4.3 and the Büchi automaton from Figure 4.8(c).

$F_A = \{(c_1, s_2), (c_2, s_2), (c_3, s_2), (c_4, s_2), (c_5, s_2), (c_6, s_2)\}$. The transition relation δ_A is shown by arrows in Figure 5.11. Note that c_4 is the only state of T_S with output y_1, and according to the definition of δ_A, state (c_4, s_1) can transit to three of the final states, (c_2, s_2), (c_4, s_2), (c_5, s_2). Furthermore, only the final state (c_4, s_2) can be infinitely often visited.

By running Algorithm 5.2, the accepted path of A has prefix $= (c_1, s_1), (c_2, s_1), (c_4, s_1)$ and suffix $= (c_4, s_2)$. Thus, $path_{T_S} = c_1, c_2, c_4, c_4, \ldots$, i.e. the robot should move to cell c_4 and remain there. Clearly this path satisfies the LTL formula. Despite the simplicity of this example, we emphasize that the automated approach from this subsection is applicable to any pair T_S, B, and thus one can obtain robotic trajectories for any desired LTL formula. ■

Team of identical robots. If we assume a team of identical robots, we can construct a team transition system T_T as in Subsection 4.3.2 and Definition 4.3. Imposing an LTL specification for the whole team means that the robots should cooperate such that the output sequence of T_T satisfies the formula. Therefore, such a specification is called global because there are no individual parts of the formula preassigned to specific robots, in contrast with works such as [88, 182, 201], where individual LTL tasks can be given to specific robots.

Since T_T has the same set of possible observations as an individual transition system T_S, the product automaton $A_T = T_T \times B$ can be constructed as in Definition 5.1. Then, Algorithm 5.2 can be used to find a path of T_T that satisfies the specification. Finally, the path of T_T has to be projected to individual robot paths, thus obtaining the sequence of cells that each robot should traverse. However, since T_T is a synchronous product of transition systems, in order to generate the desired sequence of observations while moving the team, the robots should synchronize when changing cells [112]. This synchronization is usually performed by waiting at the border of cells and ensuring that all robots cross at the same moment to their next cell - thus, the observed team behavior is exactly the sequence of states of T_T.

While this formal extension to a team of identical robots is straightforward once having the algorithm for a single robot, the practical implementation may yield difficulties due to the state space explosion problem. Thus, for $|R|$ robots, the product automaton $A_T = T_T \times B$ has $|C|^{|R|} \cdot |S|$ states, and this exponential increase results in computational problems for computing, storing, and finding an accepted path for A_T. Usually, this extension is applicable for a reduced team size (at most 3 or 4 robots) [112, 118], while for multiple robots it is computationally feasible to use the Petri net modeling formalism, as will be described in Chapter 6.

5.7 Collision avoidance using initial delay

5.7.1 Problem description

In this section we consider again a team transition system T_T as in Subsection 4.3.2 and Definition 4.3. However, we assume that in the environment a region denoted d exists corresponding to a depot where the robots are initially placed and where they will return after moving through nodes from C. Without loss of generality, we assume that this depot is a big region that is surrounding the partitioned environment (see Figure 5.12 for an example). Therefore, $C = C \cup \{d\}$ and the product automaton T_T is constructed accordingly.

The depot d should be regarded as an area where the robots are handled, e.g. stored, charged, or programmed, by a central unit. While the robots are in depot d, they cannot collide. However, once the robots leave d and evolve in regions from C, they move independently, without communicating with other robots or with the central unit. Therefore, it is possible that robots collide in nodes from C. For example, in the environment of Figure 5.12, if robot r_1 follows the sequence $dc_2c_4c_6d$ and robot r_2 follows the sequence $dc_3c_4c_5d$ then they can collide in region c_5. This section presents a technique to avoid such collisions by imposing initial delays in robot movements.

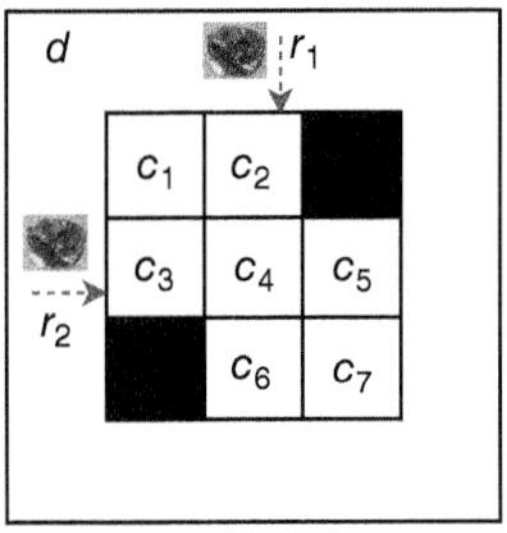

Figure 5.12 A simple environment surrounded by a depot region. A simple environment partitioned using the polytopal decomposition Algorithm 3.3 presented in Section 3.2.4 and surrounded by a depot region d in which the robots cannot collide.

We assume that a pre-computed set of trajectories $\mathcal{T}_i = \{s_{(i,1)}, s_{(i,2)}, \dots, s_{(i,|\mathcal{T}_i|)}\}$ is given such that each $s_{(i,j)}$ is a sequence of cells (nodes) of C that starts and finishes in d, i.e. $s_{(i,j)} = dc^1_{(i,j)}c^2_{(i,j)} \cdots c^{n_{i,j}}_{(i,j)}d$, with $c^k_{(i,j)} \in C, \forall k = 1, \dots, n_{i,j}$. If each robot r_i follows one trajectory from its set $\mathcal{T}_i$, the task assigned to the team is accomplished. Basically, each robot should leave depot d by entering in a node from C, moves through nodes from C, and returns to d.

Let $c^k_{(i,j)}$ be the kth cell (excluding depot) in the jth trajectory of robot r_i. We associate a crossing time $\tau^k_{i,j}$ with $c^k_{(i,j)}$ to describe the time needed by r_i from entering region $c^k_{(i,j)}$ until leaving it.

Note that if the same cell from C belongs to multiple trajectories, its associated crossing times may be different, which in fact is in accordance with a model build by partitioning the robotic environment. For example, assume that place $c \in C$ appears in trajectories j and j' of robot r_i: $c = c^k_{(i,j)} = c^{k'}_{(i,j')}$. In the jth trajectory, c is crossed by entering the region from a previous region $c^{k-1}_{(i,j)}$ and by moving towards the next region $c^{k+1}_{(i,j)}$, e.g. by following a line segment or by applying a specific control law. The moving time $\tau^k_{i,j}$ is therefore related to this local continuous trajectory, and it may be different from $\tau^{k'}_{i,j'}$, since in the j'th trajectory region p may be reached (left) from (to) other regions by different local control laws and trajectories. On the same idea, if robots have different speeds, this information is also encapsulated by the crossing times related to each tuple robot-trajectory place.

There is no crossing time associated with d, meaning that any robot r_i can immediately access C. If all robots start to move at the initial time 0, it is possible that some robots collide, because their trajectories may intersect. We assume that two robots collide if they are in the same region from C at the same time or if they swap two adjacent regions (one moves from c to c' while the other moves from c' to c). Clearly, collisions depend on the intersections of trajectories of different robots and on the crossing times for regions along these trajectories.

Recall that the robots move independently and therefore they cannot pause their movement in order to avoid collisions and yield crossing priorities to other robots. Even if such pauses were possible, deadlocks may appear in the case of circular waits of some robots [123]. However, since the robots are initially in the depot, where they are handled by a central unit, it is possible to delay the starting time for moving along each possible trajectory, in order to ensure that no collisions appear. If robot r_i follows trajectory $s_{(i,j)}$, let us denote by $\delta_{i,j}$ the time delay when r_i starts to move.

Example 5.4 *Consider the graph environment from Figure 5.12, in which C includes seven places and the depot that surrounds these regions. There are two robots r_1 and r_2, each having a single trajectory:*

- $\mathcal{T}_1 = \{s_{(1,1)}\}$, $s_{(1,1)} = dc^1_{(1,1)}c^2_{(1,1)}c^3_{(1,1)}d = dc_2c_4c_6d$;

- $\mathcal{T}_2 = \{s_{(2,1)}\}$, $s_{(2,1)} = dc^1_{(2,1)}c^2_{(2,1)}c^3_{(2,1)}d = dc_3c_4c_5d$.

The following crossing times are assumed:

- *for* $s_{(1,1)}$: $\tau^1_{1,1} = 1$, $\tau^2_{1,1} = 2$, $\tau^3_{1,1} = 1$;
- *for* $s_{(2,1)}$: $\tau^1_{2,1} = 2$, $\tau^2_{2,1} = 1$, $\tau^3_{2,1} = 1$.

If the initial delays of two trajectories are zero ($\delta_{1,1} = \delta_{2,1} = 0$)*, then the arrival and departure times of* r_1 *in* $c^2_{(1,1)}$ *(region* c_4 *in the map) are* 1 *and* 3*, respectively, while the arrival and departure times of* r_2 *in* $c^2_{(2,1)}$ *(same region* c_4*) are* 2 *and* 3*, respectively. Because both robots are in the cell* c_4 *in the time interval* [2, 3]*, they can collide.*

The main idea of the techniques presented in this section is to find initial time delays $\delta_{i,j}$ such that there are no robot collisions and the robots finish their tasks in the minimum possible time. A trivial solution for enforcing these delays would be to sequentially move the robots, i.e. first move only robot r_1; when it finishes its trajectory start moving r_2, and so on. Clearly, such an approach forbids simultaneous movements and implies a long time until each robot finishes its chosen trajectory. In a resource allocation framework, nodes C can be regarded as a shared resource for the $|R|$ robots, while the trivial solution implies a mutual exclusion for accessing the set C.

Depending on the ability of the central unit to impose the trajectory of each robot r_i from set $\mathcal{T}_i$, we formulate two different problems.

Problem 5.1 *(Decentralized)* Each robot r_i chooses its trajectory from set $\mathcal{T}_i$ without informing the central unit. Find initial time delays $\delta_{i,j}$, $i = 1, \dots, |R|$, $j = 1, \dots, |\mathcal{T}_i|$, such that there are no collisions and all robots finish the movement in the shortest time.

Problem 5.2 *(Centralized)* The central unit can impose a trajectory for each robot r_i from set $\mathcal{T}_i$. For each robot r_i, $i = 1, \dots, |R|$, find the trajectory $s_{i,j} \in \mathcal{T}_i$ it should follow and its initial time delay $\delta_{i,j}$, such that there are no collisions and all robots finish the movement in the shortest time.

The common hypothesis for both problems are the sets of trajectories $\mathcal{T}_i$ and the crossing times of robots through regions $\tau^k_{i,j}$, $i = 1, \dots, |R|$, $j = 1, \dots, |\mathcal{T}_i|$, $k = 1, \dots, n_{i,j}$.

Problem 5.1 will have as outcomes a number of $\sum_{i=1}^{|R|} |\mathcal{T}_i|$ initial time delays. Once these are available, each robot r_i (randomly) chooses a trajectory $s_{(i,j)}$ from $\mathcal{T}_i$, waits for time $\delta_{i,j}$, and then starts to evolve along $s_{(i,j)}$.

Problem 5.2 will have a total of $2 \times |R|$ outcomes: the index j for trajectory and the initial time delay $\delta_{i,j}$, for each robot r_i, $i = 1, \dots, |R|$. The trajectories and initial delays imply a minimum completion time (until all robots return to depot d). Of course, the completion time yielded by solution of Problem 5.1 may be longer, since it is minimized by accounting for any possible robot-trajectory choice.

Remark 5.1 The above problems can be formulated by ignoring depot d and by assuming that each robot is initially deployed in a place from map C. The same procedure can be followed for finding initial time delays. However, in such a scenario it is possible that the above problems become infeasible (e.g. assume two robots each with a single trajectory, the trajectories consist of the same regions from C, but are followed in opposite directions by the robots). If depot d is assumed, the problems always have solutions (trivial solution in which the robots moved sequentially).

As possible real scenarios mimicked by the above formulation, one can imagine non-communicating exploring robots, air transportation systems, and automated railway systems where specific track segments and intersections correspond to nodes from C. Initial delays rather than pausing motions may be desirable in certain situations; e.g. in air transportation it is cheaper to let planes wait in their departure airport than wait in the air, while in railway systems paused trains may perturb other traffic participants.

Example 5.5 *In Example 5.4, a possible solution for avoiding collisions can be* $\delta_{1,1} = 2 + \epsilon$ *(with* ϵ *a very small value) and* $\delta_{2,1} = 0$*, a case in which* r_1 *finishes its trajectory at time* $6 + \epsilon$ *(and* r_2 *at time 4). A better solution would be* $\delta_{1,1} = 0$ *and* $\delta_{2,1} = 1 + \epsilon$*, such that the robots are back in the depot at time* $5 + \epsilon$*.*

5.7.2 Solution for Problem 5.1 (decentralized)

If a robot r_i follows trajectory $s_{(i,j)}$, its arrival time in region $c^k_{(i,j)}$ is

$$T_A(c^k_{(i,j)}) = \delta_{i,j} + \sum_{l=1}^{k-1} \tau^l_{i,j}, \tag{5.18}$$

where $\delta_{i,j}$ is the initial delay (unknown). The departure time from $c^k_{(i,j)}$ is

$$T_D(c^k_{(i,j)}) = T_A(c^k_{(i,j)}) + \tau^k_{i,j}. \tag{5.19}$$

Consider two trajectories $s_{(i,j)}$ and $s_{(\alpha,\beta)}$ of two robots r_i and r_α, respectively. Assume that the two trajectories intersect in a cell $c \in C$, $c = c^k_{(i,j)} = c^\gamma_{(\alpha,\beta)}$ (kth region from trajectory of robot r_i is identical with γth region from trajectory of robot r_α). The robots do not collide in this place if either inequality (5.20a) or (5.20b) is true:

$$\begin{cases} T_D(c^k_{(i,j)}) \leq T_A(c^\gamma_{(\alpha,\beta)}) - \epsilon, & (5.20a) \\ T_D(c^\gamma_{(\alpha,\beta)}) \leq T_A(c^k_{(i,j)}) - \epsilon. & (5.20b) \end{cases}$$

In inequalities (5.20), ϵ is a very small number. It ensures that robots r_i and r_α will never be on the same border of a region at the same time, such that they cannot collide by swapping places (if $c^{k+1}_{(i,j)} = c^{\gamma-1}_{(\alpha,\beta)}$, or $c^{k-1}_{(i,j)} = c^{\gamma+1}_{(\alpha,\beta)}$). The above inequalities mean that the robots do not collide in cell $c = c^k_{(i,j)} = c^\gamma_{(\alpha,\beta)}$ if either r_i departs c before r_α arrives there (5.20a) or vice-versa (5.20b).

The satisfaction of one inequality from (5.20) will be enforced by appropriate values for delays $\delta_{i,j}$ and $\delta_{\alpha,\beta}$. The *disjunction* from (5.20) can be transformed into a *conjunction* of inequalities (5.21), by a so-called big number method [86]. Let N be a large number, and define a binary variable $b_{ijk,\alpha\beta\gamma}$ such that $b_{ijk,\alpha\beta\gamma} = 0$ if (5.20a) holds and $b_{ijk,\alpha\beta\gamma} = 1$ if (5.20b) is true. There is no collision in $c = c^k_{(i,j)} = c^\gamma_{(\alpha,\beta)}$ if inequalities from (5.21) simultaneously hold:

$$\begin{cases} T_D(c^k_{(i,j)}) - T_A(c^\gamma_{(\alpha,\beta)}) \leq N \cdot b_{ijk,\alpha\beta\gamma} - \epsilon, \\ T_D(c^\gamma_{(\alpha,\beta)}) - T_A(c^k_{(i,j)}) \leq N \cdot (1 - b_{ijk,\alpha\beta\gamma}) - \epsilon. \end{cases} \tag{5.21}$$

Recall that in the scenario of Problem 5.1, every robot r_i will choose a trajectory to follow from its set of trajectories $\mathcal{T}_i$. Since we cannot determine in advance which trajectory will be followed by each robot, we should compute initial time delays of all trajectories so that no matter which ones are chosen, the system will be collision free. By using (5.21) for all possible intersections between trajectories of different robots, we obtain the constraints from (5.22), where the expressions of arrival and departure times are given by Eqs. (5.18) and (5.19):

$$\begin{cases} T_D(c^k_{(i,j)}) - T_A(c^\gamma_{(\alpha,\beta)}) \leq N \cdot b_{ijk,\alpha\beta\gamma} - \epsilon, \\ T_D(c^\gamma_{(\alpha,\beta)}) - T_A(c^k_{(i,j)}) \leq N \cdot (1 - b_{ijk,\alpha\beta\gamma}) - \epsilon, \\ \qquad \forall i, \alpha \in \{i = 1, \dots, |R|\},\ i \neq \alpha,\ j \in \{1, \dots, |\mathcal{T}_i|\}, \\ \qquad \beta \in \{1, \dots, |\mathcal{T}_\alpha|\},\ k \in \{1, \dots, n_{i,j}\}, \\ \qquad \gamma \in \{1, \dots, n_{\alpha,\beta}\}, \text{s.t. } c^k_{(i,j)} = c^\gamma_{(\alpha,\beta)}. \end{cases} \tag{5.22}$$

In order to solve Problem 5.1, we have to minimize the completion time when all robots are returned to the depot, i.e.

$$\min \left(\max_{i,j} \quad T_A(c^{n_{i,j}}_{(i,j)}) \right) \tag{5.23}$$

The min-max problem (5.23) with constraints (5.22) can be transformed into the standard MILP formulation (5.24) by introducing an auxiliary variable y for the completion time.

$$\min(y)$$
$$\text{s.t.}\begin{cases} T_A(c^{n_{i,j}}_{(i,j)}) \le y \\ \text{constraints (5.22).}\end{cases} \tag{5.24}$$

MILP (5.24) can be solved by using optimization software packages [150]. It has $1 + \sum_{i=1}^{|R|} |\mathcal{T}_i|$ free (real) variables (completion time and initial time delays) and some binary variables $b_{ijk,\alpha\beta\gamma}$ (their number depending on how many intersections exist between trajectories of different robots).

Example 5.6 *The MILP (5.24) corresponding to the system in Example 5.4 is*

$$\min(y)$$
$$s.t.\begin{cases} \delta_{1,1}+4 \le y \\ \delta_{2,1}+4 \le y \\ (\delta_{1,1}+3)-(\delta_{2,1}+2) \le N\cdot b_{112,212}-\epsilon \\ (\delta_{2,1}+3)-(\delta_{1,1}+1) \le N\cdot(1-b_{112,212})-\epsilon \\ 0 \le \delta_{1,1},\delta_{2,1} \\ b_{112,212} \in \{0,1\}. \end{cases} \tag{5.25}$$

The solution of the MILP is $\delta_{1,1}=0$, $\delta_{2,1}=1+\epsilon$, $b_{112,121}=0$, and $y=5+\epsilon$.

5.7.3 Solution for Problem 5.2 (centralized)

We extend the solution from Subsection 5.7.2 to the case when the trajectory of each robot is chosen by a central unit, not by the robot. In order to implement this policy, we introduce binary variables $x_{i,j}$ for the jth trajectory of the ith robot. If $x_{i,j}=1$, then r_i will follow $s_{(i,j)}$; otherwise, $s_{(i,j)}$ will not be followed by r_i. Therefore, $\sum_{j=1}^{|\mathcal{T}_i|} x_{i,j}=1$, $\forall i=1,\ldots,|R|$. Constraints from optimization (5.24) become (5.26), where the constraints (5.26a) consider only completion time for the chosen trajectories:

$$\begin{cases} T_A(c^{n_{i,j}}_{(i,j)}) - y \le N\cdot(1-x_{i,j}) & (a) \\ \sum_{j=1}^{|\mathcal{T}_i|} x_{i,j} = 1 & (b) \\ T_D(c^k_{(i,j)}) - T_A(c^\gamma_{(\alpha,\beta)}) \le N\cdot b_{ijk,\alpha\beta\gamma}-\epsilon & (c) \\ T_D(c^\gamma_{(\alpha,\beta)}) - T_A(c^k_{(i,j)}) \le N\cdot(1-b_{ijk,\alpha\beta\gamma})-\epsilon & (d) \\ \quad \forall i,\alpha\in\{i=1,\ldots,|R|\},\ i\ne\alpha,\ j\in\{1,\ldots,|\mathcal{T}_i|\}, \\ \quad \beta\in\{1,\ldots,|\mathcal{T}_\alpha|\},\ k\in\{1,\ldots,n_{i,j}\}, \\ \quad \gamma\in\{1,\ldots,n_{\alpha,\beta}\}, \text{s.t. } c^k_{(i,j)} = c^\gamma_{(\alpha,\beta)}. \end{cases} \tag{5.26}$$

Constraints (5.26) basically mean:

1) central unit chooses one trajectory for each robot by the controller, via variables $x_{i,j}$;
2) all possible trajectories are conflict free (as in (5.24)), not only the chosen ones.

The second item can be relaxed by guaranteeing conflict free movements only for the chosen trajectories. The constraints (5.26c) and (5.26d) are modified based on variables $x_{i,j}$ and (5.27) is obtained:

$$\begin{cases} T_A(c^{n_{i,j}}_{(i,j)}) - y \leq N \cdot (1 - x_{i,j}) & (a) \\ \sum_{j=1}^{|\mathcal{T}_i|} x_{i,j} = 1 & (b) \\ T_D(c^k_{(i,j)}) - T_A(c^\gamma_{(\alpha,\beta)}) \leq N \cdot b_{ijk,\alpha\beta\gamma} + \\ \quad +N \cdot (1 - x_{i,j}) + \\ \quad +N \cdot (1 - x_{\alpha,\beta}) - \epsilon \cdot x_{i,j} & (c) \\ T_D(c^\gamma_{(\alpha,\beta)}) - T_A(c^k_{(i,j)}) \leq N \cdot (1 - b_{ijk,\alpha\beta\gamma}) + \\ \quad +N \cdot (1 - x_{i,j}) + \\ \quad +N \cdot (1 - x_{\alpha,\beta}) - \epsilon \cdot x_{\alpha,\beta} & (d) \\ \quad \forall i, \alpha \in \{i = 1, \dots, |R|\},\ i \neq \alpha,\ j \in \{1, \dots, |\mathcal{T}_i|\}, \\ \quad \beta \in \{1, \dots, |\mathcal{T}_\alpha|\},\ k \in \{1, \dots, n_{i,j}\}, \\ \quad \gamma \in \{1, \dots, n_{\alpha,\beta}\}, \text{s.t. } c^k_{(i,j)} = c^\gamma_{(\alpha,\beta)}. \end{cases} \tag{5.27}$$

Under minimization of y, each set of constraints (5.26) or (5.27) yield an MILP optimization that solves Problem 5.2. Notice that the number of constraints of MILP (5.27) is bigger than the number of constraints of MILP (5.26). However, according to the statistical analysis in [210], the computational time is smaller.

Example 5.7 *Let us compare the solutions from Subsections 5.7.2 and 5.7.3 by using five robots moving in a map sketched in Figure 3.8, which includes 100 regions (further denoted using indices in form $c_{(row,column)}$) and a depot d. Robots can move from one region to another one following directions up, down, left, and right. For example, a robot can move from $c_{(3,4)}$ to $c_{(3,5)}$, but never to $c_{(4,5)}$. Each robot should enter in the environment from the external depot d in a specific region, should reach a final region, and then leave to d. Let us assume the following requirements for the robots:*

- r_1: *enters in* $c_{(1,1)}$ *and reaches* $c_{(10,10)}$,
- r_2: *enters in* $c_{(9,1)}$ *and reaches* $c_{(1,10)}$,
- r_3: *enters in* $c_{(5,1)}$ *and reaches* $c_{(5,10)}$,
- r_4: *enters in* $c_{(1,5)}$ *and reaches* $c_{(10,5)}$,
- r_5: *enters in* $c_{(3,1)}$ *and reaches* $c_{(7,10)}$.

Let us consider first only one trajectory for each robot, trajectories that are shown in Figure 3.8. We assume that the time of crossing a region is 1 time unit

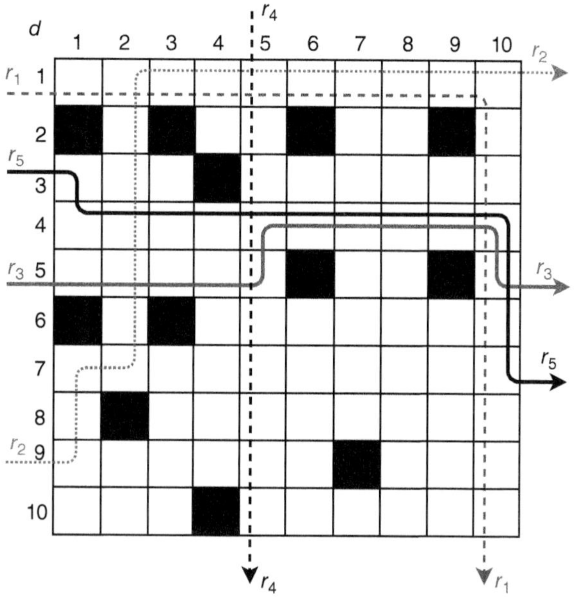

Figure 5.13 A more complex environment surrounded by a depot region. A more complex environment partitioned using the polytopal decomposition Algorithm 3.3 presented in Section 3.2.4 and surrounded by a depot region *d* in which the robots cannot collide.

if the robot enters in the region from left (right) and leaves to right (left) or enters from top (bottom) and leaves to bottom (top). Otherwise, we assume a smaller duration equal to $\sqrt{2}/2$ *time unit. MILP (5.24) has been solved in 0.0088 seconds on a computer with Intel 2.3GHz CPU and the following delays for the robot trajectories are obtained:* $\delta_{1,1} = 0$, $\delta_{2,1} = 0.2959$, $\delta_{3,1} = 5.7111$, $\delta_{4,1} = 4.7081$, *and* $\delta_{5,1} = 4.2949$, *respectively. Notice that, if* r_3 *and* r_5 *do not wait and start moving at time 0 then they will meet in region* $c_{(4,5)}$ *and could collide during the time interval* $[4.4707, 5.4142]$. *Without delay,* r_3 *goes through regions* $c_{(5,1)}c_{(5,2)}c_{(5,3)}c_{(5,4)}c_{(5,5)}$ *and reaches* $c_{(4,5)}$ *in* $1 + 1 + 1 + 1 + \sqrt{2}/2 = 4.7071$ *time units. It will leave* $c_{(4,5)}$ *at time* $4.7071 + \sqrt{2}/2 = 5.4142$. *The robot* r_5 *would arrive at* $c_{(4,5)}$ *at 4.4142 and would leave at 5.4142. Therefore, during the time interval* $[4.4707, 5.4142]$ *both* r_3 *and* r_5 *are in* $p_{(4,5)}$ *and they could collide. If both robots are delayed according to the solution of MILP (5.24) the collision will not be possible any more.*

5.8 Conclusions

In this section some path planning problems were solved by using transition system models. The main idea is to consider the discrete model of one robot that is obtained after applying a cell decomposition approach. The model is used to find a path by applying a search on the graph algorithm or, in the case of LTL formula, a new product automaton is first computed and after that graph searches are applied. In the case of a team of identical robots and a global LTL task, the model of the team is obtained by doing the synchronous product of a number of identical automata equal to the number of robots, and then the exact strategy as for a single robot is applied. This chapter presents algorithms to automatically plan the robot motion for: reachability specification for one robot, LTL formula for one robot, and an extension to a few identical robots. For reachability specifications, different graph weights and waypoint optimizations are discussed, as well as a receding horizon approach. The main drawback of the methods based on transition system models is the exponential increase in the number of states of the model when the number of robots is growing. To overcome the prohibitive computational complexity for more robots, the next chapter presents approaches based on Petri net models.

6

Path and Task Planning Using Petri Net Models

6.1 Introduction

For a standard navigation problem in multi-robot systems, where the goal consists in reaching some destination points in a shared environment, many solutions have been proposed in the literature by the Artificial Intelligence community [145, 146, 187, 188] and the Discrete Event Systems community [47, 180, 216, 217, 223]. The main issue in these works is to compute collision-free paths. In general, they combine task planning with motion control and the solution is based on some graphs obtained by discretization of the environment, as given in this book. However, the problem studied in this book assumes high-level specifications; hence the final goals (destinations) are not a priori known and should be computed together with the robot trajectories.

One possibility used to tackle the state space explosion problem of the transition system approach described in Chapter 5 is the so-called Simultaneous Task Allocation and Planning [182], based on a decomposition of the mission as in [201]. In particular, the global LTL formula is decomposed in a number of local subformulas that can be executed in parallel. Furthermore, the planning algorithm uses a specific team model, in general with less numbers of states than the full model obtained by the product of R automatons, consisting in a number of R models in parallel, interconnected through special transitions allowing distribution of the task. There are two possible limitations of this approach: (i) not all LTL formulas can be divided into subformulas and (ii) the approach in [182] is developed for co-safe LTL formulas, while the approach in this chapter considers more expressive LTL formulas.

As in Section 5.1, assume a rectangular environment $Ev \subset \mathbb{R}^2$ defined as the Cartesian product of $[0, x_{max}] \times [0, y_{max}]$, where $x_{max}, y_{max} \in \mathbb{R}$. Ev is cluttered with some convex regions of interest and is partitioned by using a cell decomposition algorithm that also decomposes these regions, as in Section 3.3. This chapter focuses on planning problems for a team of robots modeled with a Robot

Path Planning of Cooperative Mobile Robots Using Discrete Event Models, First Edition.
Cristian Mahulea, Marius Kloetzer, and Ramón González.

Motion Petri net (RMPN) $Q = \langle \mathcal{N}, \boldsymbol{m}_0, H, h \rangle$ as in Definition 4.4. The team comprises identical omnidirectional and point robots, and the mission requirement expressed a global specification for the whole team, without imposing parts of the specifications to be fulfilled by specific robots.

To better motivate the formal solutions from this chapter, we begin the presentation by an example conducted on RMTool. Assume two robots deployed in the environment from Figure 6.1, and a specification requiring that the blue and green regions are visited along trajectories, the red region is avoided, and when the robots stop at least one of them should be in the cyan region. The mission could be expressed as an LTL formula and then the approach from Section 5.6 could be used. However, LTL is a too expressive formalism for this scenario, and its corresponding algorithm can quickly become computationally intractable if the team size is increased, as mentioned in Chapter 5. Therefore, as will be detailed in Section 6.2, we propose a Boolean-based formalism for expressing such tasks, where we can impose different requirements that should take place along trajectories, i.e. while the robots are moving, and in the final state, i.e. when the robots stop. The Boolean-based formula visible in the corresponding box of RMTool from Figure 6.1 encodes the above mission. The robot trajectories from Figure 6.1 are quickly obtained (in about 0.2 seconds) and the robots do not have to synchronize along these trajectories, as they would have to in the case of LTL formulas. The development is based on an RMPN model of the robotic team in the given environment and on mathematical programming, which casts the problem in the framework of standard optimization problems.

Clearly, some requirements that could have been expressed in LTL cannot be expressed in Boolean-based formalism - examples include infinite behaviors as surveillance of some regions (infinite visits of the regions) or specific orders in which regions should be visited along trajectories. To partially overcome this, Section 6.3 proposes mathematical problem formulations for RMPN models and specifications given as LTL formula, thus resulting in entirely different methods from the ones in Chapter 5.

Thus, in this chapter we are interested in using RMPN models and optimization problems for planning a team of robots. This new methodology leads to computational advantages for larger teams of robots compared to the previous approach in Chapter 5. PN models have been used for modeling and controlling mobile robots in the recent literature [25, 43, 130, 181, 209, 215, 219]. The modeling methodology is distinct and the models have a different significance, in our case the environment being partitioned depending on the regions of interest. Many works exist in the PN literature dealing with verification of Petri net properties with Boolean or LTL specifications [62, 63, 107, 175]. Even if some structural properties exist, it is usually necessary to explore the reachability space.

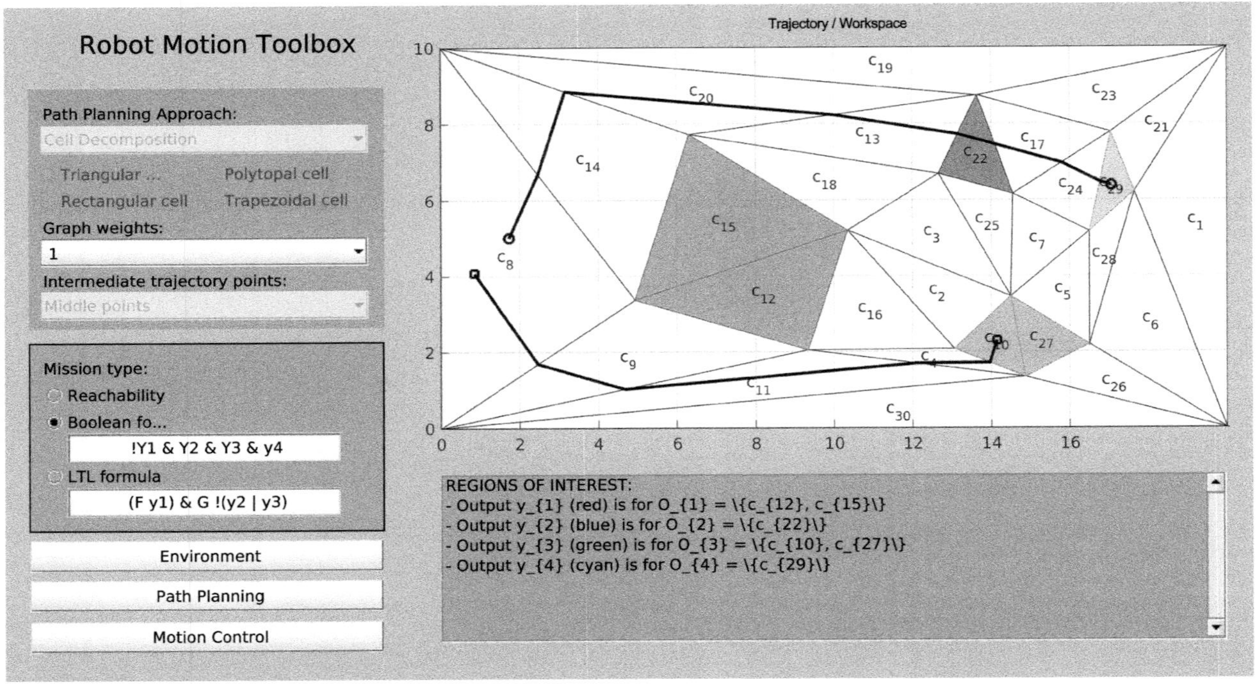

Figure 6.1 An example of a Boolean specification. The robots are required to avoid the red region (y_1), to visit the blue (y_2) and green (y_3) regions along trajectories, and when they stop, at least one of them should be in the cyan region (y_4). The mission in expressed in a Boolean-based formalism, the solution builds on a RMPN model and the shown trajectories are obtained in around 0.2 seconds, much faster than using the more general approach from Section 5.6.

The formal solution details are given in the rest of the chapter. Sections 6.2 and 6.3 are devoted to classes of specifications - Boolean-based and LTL, respectively - while Sections 6.4 and 6.5 focus on specific requirements and use the same mathematical programming ideas to obtain standard optimization formulations. Finally, Section 6.6 includes Petri net-based methods that can be used for obtaining deadlock- and collision-free robot movements along designated trajectories.

6.2 Boolean-based specifications for cooperative robots

6.2.1 Problem definition and notations

Syntactically, we assume requirements expressed as Boolean logic formulas defined over the set of variables $\mathcal{Y} = \mathcal{Y}_t \cup \mathcal{Y}_f$, where $\mathcal{Y}_t = \{Y_1, Y_2, \dots, Y_{|O|}\}$ and $\mathcal{P}_f = Y$ [1], by using the standard logical connectors $\neg$ (negation), $\wedge$ (conjunction), and $\vee$ (disjunction). The sets $\mathcal{Y}_t$ and $\mathcal{Y}_f$ refer to the same regions of interest, but the elements of $\mathcal{Y}_t$ suggest regions that should be visited (or avoided, when negated) along a trajectory, while $\mathcal{Y}_f$ suggests regions that should be visited (or avoided) in the last state of a run, as explained in the following semantics.

The specifications are interpreted over finite words by the set $H \subseteq 2^Y$, as are those generated by the RMPN system $\mathcal{Q}$ from Definition 4.4. Semantically, the lower- and upper-case notations from the above set $\mathcal{Y}$ have the following meaning when interpreted over the word generated by a run $\rho = \boldsymbol{m}_0[t_{j_1}\rangle \boldsymbol{m}_1[t_{j_2}\rangle \boldsymbol{m}_2[t_{j_3} \dots t_{j_{|r|}}\rangle \boldsymbol{m}_{|r|}$:

- $Y_i \in \mathcal{Y}_t$ evaluates to *True* over word $h(r)$ if and only if $\exists j \in \{0, 1, \dots, |r|\}$ such that $y_i \in ||\boldsymbol{V} \cdot \boldsymbol{m}_j||$;
- $y_i \in \mathcal{Y}_f$ evaluates to *True* over word $h(r)$ if and only if $y_i \in ||\boldsymbol{V} \cdot \boldsymbol{m}_{|r|}||$.

In other words, an upper-case variable refers to a proposition that is evaluated along the whole run, while a lower-case variable refers only to the final (terminal) marking. From now on, we will assume that any Boolean-based requirement φ over variables $\mathcal{Y}$ is expressed in a Conjunctive Normal Form (CNF), the conversion into such a form being possible for any logical expression [19, 205].

For example, a specification for mobile robots as $\varphi = (Y_1 \vee Y_2) \wedge \neg y_1 \wedge \neg Y_3$ requires that either region with output y_1 or y_2 is visited along the run, y_3 is always avoided, and the region with output y_1 is not true (no robot occupies it) in the final state, i.e. when all robots stop. A specification such as $\varphi = Y_1 \wedge Y_2$ requires that

1 Recall that $O = \{O_1, O_2, \dots, O_{|O|}\}$ is the set of all regions of interest (Section 4.2) and set $Y = \{y_1, y_2, \dots, y_{|O|}\}$ contains symbols (outputs) corresponding to regions of interest (Section 4.3).

regions with outputs y_1 and y_2 are visited along robot trajectories. An implication formula such as $\varphi = \neg Y_1 \vee Y_2$ is not interpreted in the intuitive sense that a visit to y_1 implies a further visit to y_2, but it is interpreted over the entire trajectories (e.g. the task is accomplished if a robot visited y_2 at a moment, even if y_1 was visited after). One cannot impose a specific order or simultaneity when visiting y_1 and y_2, as is possible when using more complex specification formalisms or robot-specific tasks [88, 118]. For more than one robot, the specification imposes a global requirement on the attainment or avoidance of regions, without allowing individual requirements such as visiting two disjoint regions with the same agent. However, this lack of expressivity, together with the RMPN model, will yield solutions whose complexity is independent of the number of robots.

The *problem* that is solved in this subsection can be formulated as follows.

Consider a team of $|R|$ identical mobile robots evolving in an environment where regions of interest labeled with elements from set Y are defined. Given a Boolean-based specification φ over set $\mathcal{Y}$ for the team, plan the robotic motion such that the resulting trajectories satisfy φ.

Assumptions. The team is represented by an RMPN system $\mathcal{Q}$ having the form from Definition 4.4. Under the natural assumption of a connected environment, the RMPN system $\mathcal{Q}$ is strongly connected, i.e. $\forall x_i, x_j \in P \cup T$, and there exists a path starting in x_i and ending in x_j. Thus, the RMPN has no spurious markings and the set of its reachable markings can be characterized by the state equation (4.4).

Let us assume that the requirement φ (expressed in CNF) consists of a conjunction of n terms: $\varphi = \varphi_1 \wedge \varphi_2 \wedge \ldots \wedge \varphi_n$. Each term φ_i, $i = 1, \ldots, n$, is a disjunction of n_i variables (negated or not) from set $\mathcal{Y}$ from the beginning of Section 6.2, having the form $\varphi_i = [Y_{j_1} \mid \neg Y_{j_1}] \vee [y_{j_1} \mid \neg y_{j_1}] \vee [Y_{j_2} \mid \neg Y_{j_2}] \vee [y_{j_2} \mid \neg y_{j_2}] \vee \ldots \vee [Y_{j_{n_i}} \mid \neg Y_{j_{n_i}}] \vee [y_{j_{n_i}} \mid \neg y_{j_{n_i}}]$. In the expression of φ_i, the square brackets "[…]" contain optional appearing terms, while "|" denotes a choice between two variables.

Solution main steps. The proposed solution begins by converting specification φ into linear restrictions over a set of $2 \cdot |Y|$ binary variables (Section 6.2.2). Then links between these binary variables and proposition satisfactions are enforced by using linear inequalities based on the RMPN system $\mathcal{Q}$. This will yield a solution for our problem based on an MILP formulation and an algorithmic translation of MILP outcome to robot trajectories (sequences of firings in the RMPN model). The MILP objective function aims to decrease the total distance traveled by robots and the amount of possible congestion, when more robots can meet in the same partition cell. For simplicity, we first handle final state requirements, i.e. formulas over $\mathcal{Y}_f$ (Section 6.2.3), and then we present the general case of trajectory requirements (Section 6.2.4). Section 6.2.5 further discusses the presented solutions. Due to the abstract model and the definition of weighting $\boldsymbol{w}$ from Algorithm 4.2, the optimality from Sections 6.2.3 and 6.2.4 does not refer to minimizing the actual traveled distance, but to minimizing a cost function that includes the expected trajectory length.

6.2.2 Linear restrictions for Boolean-based specifications

Definition 6.1 Let $\boldsymbol{x}$ be a binary vector of dimension $2 \cdot |Y|$ variables, denoted by $\boldsymbol{x} = [x_{Y_1}, x_{Y_2}, \dots, x_{Y_{|Y|}}, x_{y_1}, x_{y_2}, \dots, x_{y_{|Y|}}]^T \in \{0,1\}^{2\cdot|Y|}$, with the following interpretation:

- $x_{Y_i} = 1$ (or $\boldsymbol{x}[Y_i] = 1$) if proposition Y_i evaluates to *True* (i.e. region labeled with Y_i is visited along the team trajectory) and $x_{Y_i} = 0$ (or $\boldsymbol{x}[Y_i] = 0$) otherwise;
- $x_{y_i} = 1$ (or $\boldsymbol{x}[y_i] = 1$) if proposition y_i evaluates to *True* (i.e. a robot stops inside the region labeled with Y_i) and $x_{y_i} = 0$ (or $\boldsymbol{x}[y_i] = 0$) otherwise, $\forall i = 1, \dots, |Y|$.

Under these evaluations, the satisfaction of the imposed specification φ is equivalent to a set of n linear inequalities, each such restriction corresponding to a disjunctive term φ_i, $i = 1, \dots, n$. To formally construct these inequalities, for each φ_i, $i = 1, \dots, n$, we define a function $\alpha_i : \mathcal{Y} \to \{-1, 0, 1\}$ showing what variables from $\mathcal{Y}$ appear in disjunction φ_i and which of them are negated:

$$\alpha_i(\gamma) = \begin{cases} -1, & \text{if } \neg\gamma \text{ appears in } \varphi_i \\ 0, & \text{if } \gamma \text{ does not appear in } \varphi_i \\ 1, & \text{if } \gamma \text{ appears in } \varphi_i. \end{cases} , \forall \gamma \in \mathcal{Y} \tag{6.1}$$

The linear inequality corresponding to disjunction φ_i is given by

$$\sum_{\gamma \in \mathcal{Y}} (\alpha_i(\gamma) \cdot x_\gamma) \geq 1 + \sum_{\gamma \in \mathcal{Y}} min\left(\alpha_i(\gamma), 0\right) , \tag{6.2}$$

where $min\left(\alpha_i(\gamma), 0\right)$ is the minimum value between $\alpha_i(\gamma)$ and 0.

Informally, Eqs. (6.1) and (6.2) come from the following ideas: if the region corresponding to symbol $\gamma \in \mathcal{Y}$ is not captured in φ_i, then its corresponding binary variable is unconstrained (it has coefficient $\alpha_i(\gamma)$ equal to zero). From all regions that appear non-negated in disjunction φ_i, at least one should be visited and thus the sum of all their corresponding binary variables should be greater or equal than 1. In Eq. (6.2), the non-negated symbols have coefficient 1 and they do not alter the right-hand term, since Eq. (6.1) evaluates to 1 for these symbols. A negated symbol γ means the avoidance of a region (either along the trajectory or in the final state), which implies that its corresponding binary variable x_γ should be 0. Equivalently, $1 - x_\gamma = 1$ and because x_γ is binary we can write $1 - x_\gamma \geq 1$. The first term "1" from here is placed in the right-hand term of Eq. (6.2) via function $min\left(\alpha_i(\gamma), 0\right)$.

For a better understanding we include here several examples of applying expression (6.2) to some disjunctions:

- the inequality corresponding to y is $x_y \geq 1$, which can be satisfied if and only if the binary x_y has value 1,
- the inequality corresponding to $\neg y$ is $-x_y \geq 0$, which can be satisfied if and only if the binary x_y has value 0,
- the inequality corresponding to $Y_1 \vee y_1 \vee \neg Y_2$ is $x_{Y_1} + x_{y_1} - x_{Y_2} \geq 0$, which holds only for those binary values of $x_{Y_1}, x_{y_1}, x_{Y_2}$ for which the disjunction is *True*.

The CNF specification $\varphi = \varphi_1 \wedge \varphi_2 \wedge \ldots \wedge \varphi_n$ is algorithmically converted (by using Eq. (6.2)) into a system of n linear inequalities, one for each disjunctive term. For example, the specification mentioned in Subsection 6.2.1, $\varphi = (Y_1 \vee Y_2) \wedge \neg y_1 \wedge \neg Y_3$, translates to the following system:

$$\begin{cases} x_{Y_1} + x_{Y_2} \geq 1 \\ x_{y_1} \leq 0 \\ x_{Y_3} \leq 0. \end{cases} \tag{6.3}$$

The obtained inequalities simultaneously hold only for binary values of $\boldsymbol{x}$ for which φ evaluates to *True*, under the links given in Definition 6.1. In the following subsections we enforce these links between binary variables and proposition satisfactions by using markings of the RMPN system $\mathcal{Q}$.

6.2.3 Solution for constraints on the final state

When finding a solution for the proposed problem, one can consider various performance measures for the resulting robot movements. In the current formulation, we aim to reduce

a) the total expected distance traveled by agents and
b) the number of situations in which robots can collide.

For point (a), we weight the fired transitions with average distances for moving a robot between two adjacent cells, i.e. we aim to minimize $\boldsymbol{w}^T \cdot \sigma$, with $\boldsymbol{w}$ computed in Algorithm 4.2. For point (b), we note that, for a given firing count vector σ, the elements of vector $\boldsymbol{Post} \cdot \sigma$ contain the cumulative number of tokens from each place of RMPN induced by firings of transitions from σ. Thus, $\boldsymbol{Post} \cdot \sigma$ gives the number of visits (not necessarily at the same time moment) in partition cells, and by reducing these values we reduce the possibilities of having more robots in the same cell. We combine goals (a) and (b) as the cost function

$$\lambda \cdot \boldsymbol{w}^T \cdot \sigma + \mu \cdot \| \boldsymbol{Post} \cdot \sigma \|_\infty,$$

where λ and μ are design parameters and $\| . \|_\infty$ denotes the maximum norm of a vector. For obtaining a linear cost function, we minimize

$$\lambda \cdot \boldsymbol{w}^T \cdot \sigma + \mu \cdot b,$$

where b upper bounds any element of $\boldsymbol{Post} \cdot \sigma$. The above considerations together with the goal of obtaining a final marking at which the formula is satisfied are captured by the following MILP:

$$
\begin{aligned}
&\min \lambda \cdot \boldsymbol{w}^T \cdot \sigma + \mu \cdot b \\
\text{s.t. } &\boldsymbol{m} = \boldsymbol{m}_0 + \boldsymbol{C} \cdot \sigma \\
&\textstyle\sum_{\gamma \in \mathcal{Y}_f} \left(\alpha_i(\gamma) \cdot x_\gamma\right) \geq 1 + \sum_{\gamma \in \mathcal{Y}_f} min\left(\alpha_i(\gamma), 0\right), \forall \varphi_i \\
&N \cdot x_\gamma \geq \boldsymbol{v}_\gamma \cdot \boldsymbol{m}, \forall \gamma \in \mathcal{Y}_f \\
&x_\gamma \leq \boldsymbol{v}_\gamma \cdot \boldsymbol{m}, \forall \gamma \in \mathcal{Y}_f \\
&\boldsymbol{Post} \cdot \sigma \leq b \cdot \boldsymbol{1}^T \\
&\boldsymbol{m} \in \mathbb{N}_{\geq 0}^{|P|}, \sigma \in \mathbb{N}_{\geq 0}^{|T|}, \boldsymbol{x} \in \{0\}^{|Y|} \times \{0,1\}^{|Y|}, b \geq 0.
\end{aligned}
\tag{6.4}
$$

In (6.4), $\boldsymbol{v}_\gamma$ is the characteristic vector of $\gamma \in \mathcal{Y}_f$ and the first $|Y|$ binary variables from $\boldsymbol{x}$ (for trajectory requirements) are set to zero, since specifications from this subsection do not include such constraints. MILP (6.4) has $(2 \cdot |Y| + n + 2 \cdot |P|)$ constraints and $(|P| + |T| + |Y| + 1)$ unknowns, from which $|Y|$ variables are binary.

The second set of constraints from (6.4) links the formula conjunctions to binary variables for final regions. If the final region γ is not captured in φ_i, then its corresponding binary variable is unconstrained (coefficient $\alpha_i(\gamma)$ is zero). Regions that appear non-negated or negated in disjunction φ_i yield (through (6.1)) coefficients "+1" or "-1", respectively, in the left-hand term, and the negated regions also decrease the value of the right-hand close parenthesis. For example, if $\varphi_i = y_1 \vee y_2$, at least one of the two regions should be visited such that $x_{y_1} + x_{y_2} \geq 1$. If $\varphi_i = \neg\gamma$, then x_γ should be 0, i.e. $1 - x_\gamma = 1$, and since x_γ is binary we can write $1 - x_\gamma \geq 1$; the first "1" from here is placed in the right-hand term via function $min\left(\alpha_i(\gamma), 0\right)$.

The third and fourth constraints from (6.4) enforce the correct values of binary variables x_{y_i} corresponding to observations in final positions. Recall that $|R|$ is the number of robots (tokens of Q) and $N \geq |R|$ should be considered.

As an alternative cost function for MILP (6.4), it is possible to minimize the number of transitions (robot movements) along the team trajectory, by choosing the objective function $\boldsymbol{1}^T \cdot \sigma$.

Based on the optimal solution σ of (6.4), the robot (token) trajectories are obtained by firing the enabled transitions and by storing the sequence of places visited by each token. The strategy is given in Algorithm 6.1.

Lemma 6.1 *If the optimal solution σ of (6.4) satisfies $\|\boldsymbol{Post} \cdot \sigma\|_\infty = 1$ (that is equivalent to $b = 1$), then there are no collisions possible during robot movements.*

Proof: Since $\boldsymbol{Post} \cdot \sigma$ counts the number of tokens in each place corresponding to the firing vector σ, the hypothesis basically says that each partition cell is visited at most once during team movement. ■

Algorithm 6.1: Iterative construction of agent strategies

Input: $\langle P, T, \boldsymbol{C} \rangle, \boldsymbol{m}_0, \sigma$

Output: Robot movement strategies

Let $\boldsymbol{m} = \boldsymbol{m}_0$;

while $\mathbf{1}^T \cdot \sigma > 0$ **do**

  Let $t \in T$ s.t. $\sigma[t] > 0 \ \wedge \ \boldsymbol{m}[^{\bullet}t] > 0$;

  Pick any robot i in $^{\bullet}t$;

  Assign movement according to t to robot i;

  Let $\boldsymbol{m} \ := \ \boldsymbol{m} + \boldsymbol{C}[\cdot, t]$;

  Let $\sigma[t] \ := \ \sigma[t] - 1$;

Note that a path planning problem can be divided into two steps: (a) the first one (tackled by current work) is to compute mission-fulfilling trajectories for the robots (while trying to avoid the congestion); (b) second, having the trajectories, one can try to avoid collisions and deadlocks by adding an additional controller. If $\| \boldsymbol{Post} \cdot \sigma \|_{\infty} > 1$, congestion can occur in places $p \in P$ for which $(\boldsymbol{Post} \cdot \sigma)[p] > 1$, and further steps have to be taken for collision avoidance and deadlock prevention. To this goal, one can try to use specific Petri net models with *capacity* constraints on some places [123] and supervisory control theory of discrete event systems [47, 167, 178, 180, 184], one possibility being included in Section 6.6. However, there are no guarantees that a deadlock-free movement is possible for any obtained trajectories, and in such cases the procedure for generating trajectories should be altered.

Two properties of the RMPN model for the system considered here are used to guarantee the correctness of Algorithm 6.1:

- The *RMPN is a live state machine* and hence all solutions of the state equation (4.4) are reachable markings. This ensures that the marking $\boldsymbol{m}$ solution of (6.4) is a reachable marking, i.e. not a spurious one;
- Since $\boldsymbol{w} \geq 0$ (that is a natural assumption being related to distances or energy), the *paths of the robots have no cycles*. This property also ensures that σ solution of (6.4) is not a "*spurious*" vector, i.e. there exists a fireable firing sequence σ with the firing count vector σ.

6.2.4 Solution for constraints on trajectory and final state

To allow constraints on final team deployment (set $\mathcal{Y}_f$) and on team trajectory (set $\mathcal{Y}_t$), the first idea was to include constraints on the firing count vector σ in Eq. (6.4). However, due to general constraints, some robot trajectories may need cycles. When solving (6.4), these cycles would *not be included* in the obtained solutions, i.e. *spurious firing vectors* would appear. This can be observed by considering the state equation corresponding to a reachable marking $\boldsymbol{m} = \boldsymbol{m}_0 + \boldsymbol{C} \cdot \sigma$. Let us assume that σ corresponds to a firing sequence σ that contains a cycle, i.e. $\sigma = \sigma' + \sigma''$, with σ'' the cycle's firing count vector. Since in a state machine RMPN a T-semiflow is a cycle, this implies that $\boldsymbol{C} \cdot \sigma'' = 0$ [192]. Obviously, the cost function of (6.4) would yield vector σ' rather than σ as the optimal solution, so the firing sequence σ would not be obtained.

To avoid spurious firing count vectors, we consider a sequence of k markings $\boldsymbol{m}_1, \boldsymbol{m}_2, \dots, \boldsymbol{m}_k$ such that $\boldsymbol{m}_1 = \boldsymbol{m}_0 + \boldsymbol{C} \cdot \sigma_1$, $\boldsymbol{m}_0 - \boldsymbol{Pre} \cdot \sigma_1 \geq \boldsymbol{0}$; $\boldsymbol{m}_2 = \boldsymbol{m}_1 + \boldsymbol{C} \cdot \sigma_2$, $\boldsymbol{m}_1 - \boldsymbol{Pre} \cdot \sigma_2 \geq \boldsymbol{0}$; … ; $\boldsymbol{m}_k = \boldsymbol{m}_{k-1} + \boldsymbol{C} \cdot \sigma_k$, $\boldsymbol{m}_{k-1} - \boldsymbol{Pre} \cdot \sigma_k \geq \boldsymbol{0}$. Informally, these constraints enforce that between RMPN states $\boldsymbol{m}_{i-1}$ and $\boldsymbol{m}_i$ each token moves *at most* through one transition, i.e. each robot advances a maximum of one cell. This approach also simplifies the construction of agents' strategies.

Putting together the cost function concept from MILP (6.4), the RMPN state equations for the sequence of k markings, and the restrictions concerning the binary variables x_{y_i} and x_{Y_i}, the following optimization problem is obtained:

$$
\begin{aligned}
\min\ & \lambda \cdot \boldsymbol{w}^T \cdot \sum_{i=1}^{k} \sigma_i + \mu \cdot b \\
\text{s.t. } & \boldsymbol{m}_i = \boldsymbol{m}_{i-1} + \boldsymbol{C} \cdot \sigma_i, i = 1, \dots, k \\
& \boldsymbol{m}_{i-1} - \boldsymbol{Pre} \cdot \sigma_i \geq 0, i = 1, \dots, k \\
& \sum_{\gamma \in \mathcal{Y}} \left(\alpha_i(\gamma) \cdot x_\gamma\right) \geq 1 + \sum_{\gamma \in \mathcal{Y}} min\left(\alpha_i(\gamma), 0\right), \forall \varphi_i \\
& N \cdot x_\gamma \geq \boldsymbol{v}_\gamma \cdot \boldsymbol{m}_k, \forall \gamma \in \mathcal{Y}_f \\
& x_\gamma \leq \boldsymbol{v}_\gamma \cdot \boldsymbol{m}_k, \forall \gamma \in \mathcal{Y}_f \\
& N \cdot (k+1) \cdot x_\gamma \geq \boldsymbol{v}_\gamma \cdot \left(\sum_{i=0}^{k} \boldsymbol{m}_i \right), \forall \gamma \in \mathcal{Y}_t \\
& x_\gamma \leq \boldsymbol{v}_\gamma \cdot \left(\sum_{i=0}^{k} \boldsymbol{m}_i \right), \forall \gamma \in \mathcal{Y}_t \\
& \left(\boldsymbol{Post} \cdot \sum_{i=1}^{k} \sigma_i \right) \leq b \cdot \boldsymbol{1}^T \\
& \boldsymbol{m}_i \in \mathbb{N}_{\geq 0}^{|P|}, \sigma_i \in \mathbb{N}_{\geq 0}^{|T|}, i = 1, \dots, k \\
& \boldsymbol{x} \in \{0, 1\}^{|\mathcal{Y}|}, b \geq 0.
\end{aligned}
\tag{6.5}
$$

The optimization problem (6.5) is a standard MILP problem [34], for which there exists complete algorithms for obtaining the optimal solution, e.g. [102, 150]. Its solution $(\sigma_1, \sigma_2, \dots, \sigma_k)$ constitutes a sequence of firing count vectors for the RMPN model Q and is converted into robot trajectories as follows. For each σ_i, $i = 1, \dots, k$, any token moves at most through one transition, and lines 3-5 of Algorithm 6.1 indicate the moving robots.

Summing up the above details, (6.5) gives a solution for the problem formulated in Subsection 6.2.1, while the cost function accounts for the total expected distance traveled by the robots and the possible congestion in cells from the partitioned environment. The constraints of (6.5) ensure the following:

- the correct functioning of model $\mathcal{Q}$ (first two lines with constraints); in total $((2 \cdot k) \cdot |P|)$ constraints,
- the satisfaction of formula φ through its disjunctive terms and binary variables (third constraint); in total n constraints,
- the link between binary variables corresponding to the formula and RMPN markings for the final requirements (constraints 4 and 5; in total $(2 \cdot |Y|)$ constraints) and for the trajectory requirements (constraints 6 and 7; in total $(2 \cdot |Y|)$ constraints),
- upper bound b for elements of vector $\boldsymbol{Post} \cdot \sum_{i=1}^{k} \sigma_i$, for capturing its maximum norm (constraints 8; in total $|P|$ constraints),
- positivity restrictions for unknown variables $\boldsymbol{m}_i$, σ_i and b; $(k \cdot (|P| + |T|) + 1)$ constraints.

Remark 6.1 Instead of considering the second term of cost function from MILP (6.5), one could completely avoid collisions (rather than reducing congestions) by adding constraints of the form $\sigma_k[t_{i,j}] + \sigma_k[t_{j,i}] \leq 1,\ \forall i, j, k$. Such constraints would forbid two robots from adjacent cells to switch positions. However, such a team movement strategy would require synchronizations when robots change cells, in order to exactly follow the order of firings from successive firing count vectors σ_k.

Remark 6.2 The constant k in MILP (6.5) is a design parameter giving the maximum number of intermediate discrete states (markings) of each robot. The theoretical upper bound of k is $|T|$, because in the worst case scenario, a robot has to once follow each transition from RMPN (e.g. imagine a string-like RMPN where the "first" and "last" places have different outputs, a robot starts from the "first" place, and the formula is required to satisfy along the trajectory the output of the "last" place and to satisfy in the final state the output of the "first" one). However, in practice, much lower values of k suffice. When k is chosen too small, the problem (6.5) becomes unfeasible. If k is larger than needed, some intermediate firing vectors σ_i will become zero in solution of (6.5).

6.2.5 Discussion on the above solutions

Solution to use. When the Boolean-based specification φ contains only symbols from $\mathcal{Y}_f$, one should use the solution from Subsection 6.2.3, consisting in MILP (6.4) and Algorithm 6.1. In this case, the MILP (6.4) has far less constraints and unknowns than MILP (6.5).

For a general specification that also includes symbols from $\mathcal{Y}_t$, the solution from Subsection 6.2.4 (MILP (6.5)) should be used. One can start with a fairly low value for k, solve MILP (6.5), and increase k if the optimization fails to return a solution. The moving strategy for each robot is obtained by concatenating the transitions given by the obtained sequence of firing count vectors.

Robot synchronization. For both of the above solutions, the obtained trajectory of each robot basically satisfies a part of formula φ, such that the whole team accomplishes task φ. Because φ is a Boolean-based formula as described above, it cannot impose specific orderings or simultaneous visits of regions in Y. Therefore, each robot can individually follow its trajectory, without synchronizing with other team members.

Recalling the limitations of our approach - lack of expressivity for imposing orders when visiting regions and reducing the possible congestions rather than ensuring a collision-free movement with no deadlocks - we mention that robot synchronization would become necessary for specifications or for movement procedures that try to reduce such conservativeness.

Solution complexity. An MILP problem belongs to the NP-hard complexity class [60]. Usually, the computational burden is characterized by the number of unknowns and constraints. The MILP (6.5) (for the case of a general specification on trajectory and final state) has a number of

$$(k \cdot (|P| + |T|) + 2 \cdot |Y| + 1)$$

integer unknowns ($\boldsymbol{m}_i$, σ_i, $\boldsymbol{x}$, b) and a total number of

$$(k \cdot (3 \cdot |P| + |T|) + 4 \cdot |Y| + |P| + 1)$$

constraints. The number of constraints and unknowns of MILPs (6.4) and (6.5) does not depend on the team size N. Some data for the computational complexity is mentioned in the examples from Subsection 6.2.7.

6.2.6 Suboptimal solution

The optimal solution from Subsection 6.2.4 may exhibit a high computational complexity, especially when one chooses a large number k of intermediate steps for the trajectory. In this section we lower this complexity by reducing the size of the RMPN model and by solving the MILP on this reduced model.

The idea of reducing the RMPN $\mathcal{Q}$ (Definition 4.4) is to iteratively combine any places p_i and p_j from P that satisfy $\{p_j\} \in \left(p_i^{\bullet}\right)^{\bullet}$ and $h(p_i) = h(p_j)$ (i.e. any places that have the same output and are connected through a single transition). This reduction technique is synthesized in Algorithm 6.2, and the reduced RMPN model $\tilde{\mathcal{Q}}$ has the property that its output changes when a transition fires. If one thinks of the environment partition, the reduction means that any adjacent cells

Algorithm 6.2: Reduce the RMPN model by joining places with the same output

Input: $Q = \langle\langle P, T, F\rangle, \boldsymbol{m}_0, H, h\rangle$

Output: $\tilde{Q} = \langle\langle \tilde{P}, \tilde{T}, \tilde{F}\rangle, \tilde{\boldsymbol{m}}_0, H, h\rangle$

$\tilde{P} = P; \tilde{T} = T; \tilde{F} = F; \tilde{\boldsymbol{m}}_0 = \boldsymbol{m}_0$

while $\exists p_i, p_j \in \tilde{P}$ *such that* $\{p_j\} \in \left(p_i^{\bullet}\right)^{\bullet}$ *and* $h(p_i) = h(p_j)$ **do**

  Let $t_k = p_i^{\bullet} \cap {}^{\bullet}p_j$ and $t_l = {}^{\bullet}p_i \cap p_j^{\bullet}$

  $\tilde{T} = \tilde{T} \setminus \{t_k, t_l\}$

  $\tilde{F} = \tilde{F} \setminus \{(p_i, t_k), (t_k, p_j), (p_j, t_l), (t_l, p_i)\}$

  $\tilde{\boldsymbol{m}}_0[p_i] = \tilde{\boldsymbol{m}}_0[p_i] + \tilde{\boldsymbol{m}}_0[p_j]$

  $\tilde{P} = \tilde{P} \setminus \{p_j\}$

that satisfy the same region(s) of interest are collapsed into a single place. In a different formalism, such a reduced system is called a quotient of the initial system, constructed with respect to equivalence classes yielded by an observation map [155].

On the reduced RMPN system $\tilde{Q}$ we can apply the same procedure as in Subsection 6.2.4. Notice that, for a fixed value of k, the reduced size of the model induces less variables and constraints in the MILP (6.5). Moreover, the upper bound of k from Remark 6.2 is in general significantly reduced. The solution of (6.5) yields a sequence of transitions/markings on the reduced RMPN $\tilde{Q}$. In this sequence only non-empty firing vectors σ_i from (6.5) are considered (see Remark 6.2), and we denote this sequence by $\tilde{\rho} = \tilde{\boldsymbol{m}}_0[\tilde{t}_{j_1}\rangle\tilde{\boldsymbol{m}}_1[\tilde{t}_{j_2}\rangle\tilde{\boldsymbol{m}}_2[\tilde{t}_{j_3} \dots \tilde{t}_{j_k}\rangle\tilde{\boldsymbol{m}}_{\tilde{k}}$, where $\tilde{k} \leq k$ and $\tilde{\boldsymbol{m}}_i \neq \tilde{\boldsymbol{m}}_{i+1}, i = 0, 1, \dots, \tilde{k} - 1$.

The solution $\tilde{\rho}$ basically shows how the observations from $H \subseteq 2^Y$ should be changed such that φ is *True*, but it does not give agent trajectories as in the case of full system Q. The firing of a single transition in $\tilde{Q}$ corresponds to the firing of a sequence of transitions in the original Q. Therefore, we need to project $\tilde{\rho}$ to a sequence on the original RMPN model to obtain the robot motions. This projection is always possible, because the construction from Algorithm 6.2 guarantees that the team can produce the sequence of outputs from $\tilde{\rho}$, although some outputs are repeated in Q. These repetitions do not affect the satisfiability of Boolean-based φ [19].

The procedure to project the solution is iterative. We show how the first sequence corresponding to $\tilde{\boldsymbol{m}}_0[\tilde{t}_{j_1}\rangle\tilde{\boldsymbol{m}}_1$ is obtained. A linear programming problem (LPP) is solved in order to obtain a firing sequence in Q corresponding to $\tilde{t}_{j_1}$ from $\tilde{Q}$:

- Remove from Q all places (together with input and output transitions and corresponding arcs) having outputs different than the ones in $\tilde{\boldsymbol{m}}_0$ and $\tilde{\boldsymbol{m}}_1$. Formally, a place $p \in P$ is removed if $h(p) \neq ||\tilde{\boldsymbol{V}} \cdot \tilde{\boldsymbol{m}}_0||$ or $h(p) \neq ||\tilde{\boldsymbol{V}} \cdot \tilde{\boldsymbol{m}}_1||$, where $\tilde{\boldsymbol{V}}$ is the matrix of characteristic vectors of $\tilde{Q}$. Let $\langle \overline{\mathcal{N}}, \overline{\boldsymbol{m}}_{\boldsymbol{0}} \rangle$ be the resulting RMPN system. The removal of places and transitions ensures that no other output (that could violate the formula) is observed during the trajectory;

- Remove from $\langle \overline{\mathcal{N}}, \overline{\boldsymbol{m}}_{\boldsymbol{0}} \rangle$ all the strongly connected components that do not have any token in $\tilde{\boldsymbol{m}}_0$ (no transitions can be fired in such components). Thus, $\overline{\mathcal{N}}$ now contains only live strongly connected components;
- The first LPP constraint is the state equation of $\langle \overline{\mathcal{N}}, \overline{\boldsymbol{m}}_{\boldsymbol{0}} \rangle$: $\overline{\boldsymbol{m}}_f = \overline{\boldsymbol{m}}_0 + \overline{\boldsymbol{C}} \cdot \sigma$;
- The second constraint ensures that the output at $\overline{\boldsymbol{m}}_f$ is the same as the one at $\tilde{\boldsymbol{m}}_1$, i.e. $\overline{\boldsymbol{V}} \cdot \overline{\boldsymbol{m}}_f = \tilde{\boldsymbol{V}} \cdot \tilde{\boldsymbol{m}}_1$, where $\overline{\boldsymbol{V}}$ is the matrix formed by characteristic vectors of $\overline{\mathcal{N}}$;
- Solve the LPP minimizing the cost function $\boldsymbol{1}^T \cdot \sigma$. Since $\overline{\mathcal{N}}$ is a state machine composed of live strongly connected components, a LPP solved with Simplex method is guaranteed to return a feasible integer solution [192]. The solution gives the firing sequence on the original system Q (hence the runs for robots) and the marking $\overline{\boldsymbol{m}}_f$ of Q corresponding to $\tilde{\boldsymbol{m}}_1$ of $\tilde{Q}$.

The previous procedure is repeated for the second step of $\tilde{r}$ by taking $\overline{\boldsymbol{m}}_f$ as the initial marking. Thus, the projection of $\tilde{r}$ to a solution of Q is done by solving at most k LPPs.

Overall, the procedure from this section requires the reduction from Algorithm 6.2, an MILP problem with fewer variables and constraints than the one from (6.5), and a number of $\tilde{k} \leq k$ LPP problems on reduced systems of type $\overline{\mathcal{N}}$. In all the simulations we performed, this procedure required less computation time than the one from Subsection 6.2.4. However, the reduced MILP and the local minimization from the $\tilde{k}$ LPPs do not guarantee the optimality of the solution in Q, as was the case in Subsection 6.2.4. This is because $\tilde{Q}$ *looses* the number of transitions of Q that should fire so that a desired output is obtained.

6.2.7 Simulation examples

This section illustrates the usage of our method for planning a team of mobile robots, an approach that is implemented in RMTool. Our implementation includes the external MILP and LPP solvers from [102, 150]. In this case, we simply consider unitary weights $\boldsymbol{w} = \boldsymbol{1}$ and $\lambda = \mu = 1$ in cost functions of MILPs (6.4) and (6.5). Thus, in this section we refer to the total number of firing transitions as the minimized cost.

We consider the environment depicted in Figure 6.2, where ten polygonal regions are defined and represented in colored borders, for easier observation of their overlapping. Algorithm 4.2 from Section 4.4 yields the RMPN system Q as

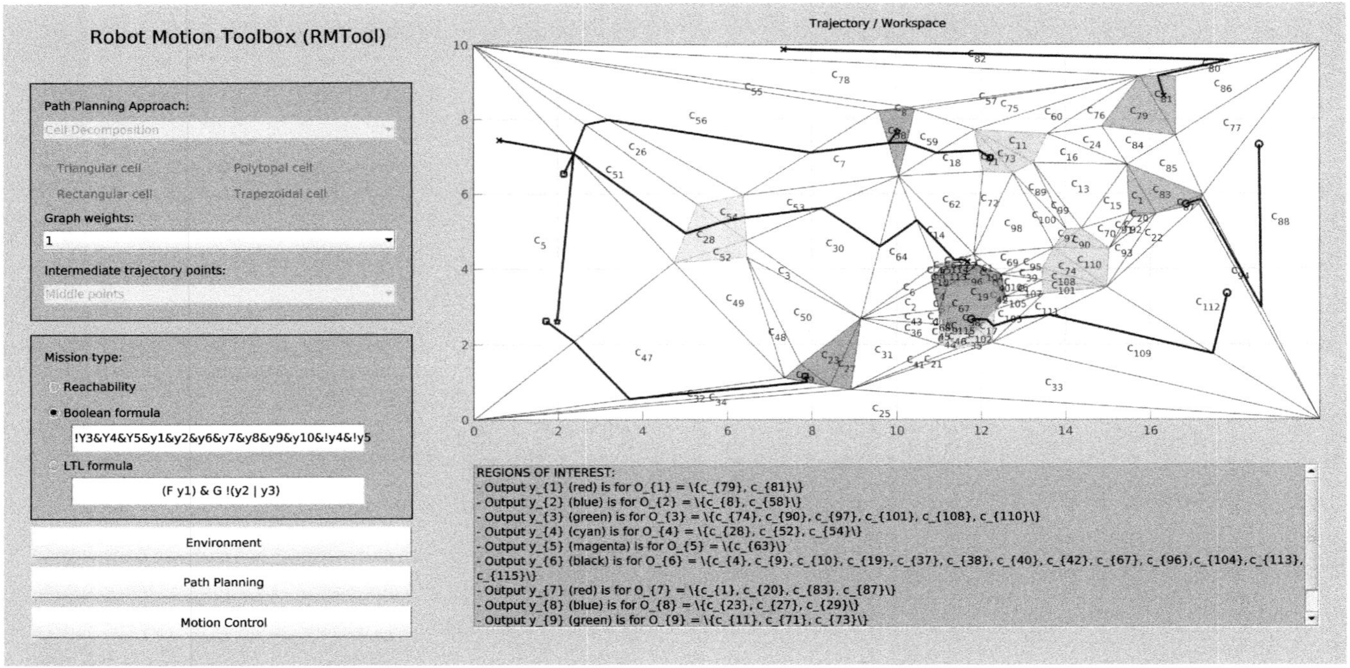

Figure 6.2 Random environment with ten regions of interest and seven robots simulated in RMTool. Environment with ten regions of interest, labeled with elements of set $Y = \{y_1, y_2, \dots, y_{10}\}$, and seven robots initially deployed in: four in c_5, and one in each of the following cells: c_{82}, c_{88} and c_{112}. Triangular partition of the environment has 116 cells. Solution (optimal with respect to the overall number of transitions) comprises a total number of 34 movements between cells. Each robot follows its trajectory and stops in the corresponding cells, and thus the team fulfills mission φ.

follows. The environment is partitioned by using a constrained triangular decomposition as in Section 3.2.3. The resulting partition has 116 cells (labeled with elements of set $C = \{c_1, c_2, \dots, c_{116}\}$) and is shown in Figure 6.2. This results in 344 transitions in T, given by adjacency between cells (two triangles are adjacent if they share an entire facet). The observation map h is easily created based on the inclusion of each cell in some regions of interest, e.g. $h(c_5) = \emptyset$, $h(c_{28}) = y_4$, $h(c_{67}) = \{y_6, y_{10}\}$. System Q has seven tokens and the initial marking is given by the initial team deployment: $\boldsymbol{m}_0[p_5] = 4$, $\boldsymbol{m}_0[p_{82}] = \boldsymbol{m}_0[p_{88}] = \boldsymbol{m}_0[p_{112}] = 1$, and $\boldsymbol{m}_0[p_i] = 0,\ \forall i \in \{1, \dots, 116\},\ i \neq 5, 82, 88{,}112$.

Considering the syntax and semantics assumed in this section, the team mission is given by the specification:

$$\varphi = \neg Y3 \wedge Y4 \wedge Y5 \wedge y1 \wedge y2 \wedge y6 \wedge y7 \wedge y8 \wedge y9 \wedge y10 \wedge \neg y4 \wedge \neg y5. \quad (6.6)$$

This means that the third region should be avoided, the fourth and fifth regions should be visited during the trajectories, but no robot should finally remain inside them, and the regions O_1, O_2, O_6, O_7, O_8, O_9, and O_{10} should be occupied when the robots stop.

By adopting the optimal solution described in Section 6.2.4 with a maximum number of steps $k = 10$, the firing sequences translate to the following runs for the robots, which can be followed without any synchronization among agents (Section 6.2.5):

$$\begin{aligned}
&\text{Robot } r_1\text{: } c_5,\ c_{47},\ c_{32},\ c_{29}\\
&\text{Robot } r_2\text{: } c_{88},\ c_{94},\ c_{87}\\
&\text{Robot } r_3\text{: } c_{82},\ c_{80},\ c_{81}\\
&\text{Robot } r_4\text{: } c_5,\ c_{51},\ c_{26},\ c_{56},\ c_7,\ c_{58}\\
&\text{Robot } r_5\text{: } c_5,\ c_{51},\ c_{26},\ c_{56},\ c_7,\ c_{58},\ c_{59},\ c_{18},\ c_{71}\\
&\text{Robot } r_6\text{: } c_{112},\ c_{109},\ c_{111},\ c_{103},\ c_{17},\ c_{38}\\
&\text{Robot } r_7\text{: } c_5,\ c_{51},\ c_{28},\ c_{54},\ c_{53},\ c_{30},\ c_{64},\ c_{14},\ c_{63},\ c_{12}.
\end{aligned} \quad (6.7)$$

The MILP problem from Section 6.2.4 includes 4621 variables, 1160 equality constraints, and 1328 inequality constraints. The solution was obtained in around 1.9 seconds on an i7-6700 CPU. Under the same conditions, if k were set to 20, the running time decreases to 1 second but the time of creating the model increases.

The actual robotic trajectories are presented in Figure 6.2, and were constructed by connecting the middle points of the common edges shared by successive cells from each robot's path. Finally, each robot converges to the centroid of the last visited cell.

As mentioned, the solution complexity is not influenced by the team size. For example, if $|R| = 20$ robots were considered for the above case, the solution is obtained in the same amount of time. More simulation results and comparisons are given in [165].

6.3 LTL specifications for cooperative robots

6.3.1 Problem definition and solution

This section considers the following problem: given an RMPN modeling a team of identical robots and a task such as the LTL formula over the set Y, find a movement strategy for the robotic team to satisfy the task.

Recall that LTL formulas were introduced at the end of Section 4.7 as being satisfiable by a finite prefix followed by an infinite repetition of a suffix. The idea of the proposed solution is to choose a sequence of observations that satisfies the LTL formula and then to generate RMPN firings that produce that sequence. Our solution is based on three main steps:

i) An accepted run ρ is chosen from the Büchi automaton (see Definition 4.9) corresponding to the LTL formula φ;
ii) For each transition of run ρ, we search a sequence of firings for the RMPN model such that the generated observations produce the chosen transition;
iii) The robots moving strategies are obtained by concatenating the sequences of firings from step (ii) and by imposing synchronization moments between these sequences.

Step (i): The Büchi automaton B corresponding to formula φ is constructed with existing software tools, e.g. [59, 74]. Before finding accepted runs, Algorithm 6.3 calls in line 1 the Algorithm 4.4. As a result, some transitions of B are removed, keeping only the one that can be generated by the inputs generated by the model Q.

Afterwards, we construct a set Γ containing a finite number of accepted runs of B. For this, we consider a small number $\kappa \in \mathbb{N}$ (usually $\kappa \leq 4$) and lines 2-11 of Algorithm 6.3. Paths with a prefix-suffix structure in B are found by running searches with the k-shortest path algorithm [220] on the adjacency graph corresponding to transitions of B. From each initial state s_0 of B we find at most κ shortest paths to each final state s_f, these paths being prefixes (lines 4-5). Note that the k-shortest path algorithm does not include cycles in the returned paths; therefore it is possible to obtain less than κ paths between two given nodes of a graph. On lines 6-10 we construct a set of paths that would bring B back to the final state s_f. The intermediate set S_{s_f} contains states that can transit to s_f. This set is needed because if one used a graph search with the same node as source and destination, one would get a path of length 1, even if that node does not have a self-loop. On line 8, "*ceil*(.)" denotes rounding toward positive infinity in order to obtain at least one suffix for each state in S_{s_f}. Each run added in set Γ (line 12) includes the prefix (driving B to a final state) and one iteration of the suffix (leading B back to that final state). Thus, each element of Γ is finite, and the infinite length needed for LTL semantics will arise from infinitely repeating the suffix and

Algorithm 6.3: Construct set Γ of accepted runs

Input: RMPN Q, Büchi B, R, κ
Output: Trimmed Büchi B, set of runs Γ
1 Reduce B by execute Algorithm 4.4;
2 $\Gamma = \emptyset$;
3 **for** $s_0 \in S_0$ *and* $s_f \in F$ **do**
4 Find at most κ paths in graph of B from s_0 to s_f (using k-shortest path algorithm);
5 Denote the set of above paths by *Pref*;
6 Let $S_{s_f} = \{s \in S | \exists \omega \in \Sigma_B \text{ with } s \xrightarrow{\omega}_B s_f\}$
7 **for** $s \in S_{s_f}$ **do**
8 Find at most *ceil* $\left(\frac{\kappa}{|S_{s_f}|}\right)$ paths from s_f to s;
9 Move s_f from the beginning of each path to the end of path;
10 Denote the set of above paths by *Suff*;
11 **if** *Pref* $\neq \emptyset$ *and Suff* $\neq \emptyset$ **then**
12 Append to each path from *Pref* each path from *Suff* and add the resulting run to set Γ;
13 For each run from Γ, store the index for beginning its suffix;

having corresponding repetitions for transitions in model Q; therefore, we store on line 13 the element of each run from where repetitions should occur.

The subsequent steps (ii) and (iii) of the method will be iterated by taking at each iteration a different run ρ from the constructed set Γ. Once a run can be followed due to RMPN's observations (step (ii) succeeds), the iteration of steps (ii) and (iii) can be stopped and the obtained solution returned - all these ideas will be algorithmically given after describing the next steps.

Let us denote the currently chosen run of B by $\rho = s_0\ s_1\ \cdots\ s_p\ \cdots\ s_{L_\rho}$, where L_ρ is the number of states from the prefix and suffix of ρ, s_p and s_{L_ρ} are the same final state to be infinitely often visited, and the suffix to be repeated is $s_{p+1}\ \ldots\ s_{L_\rho}$. If $L_\rho = p + 1$, then the suffix has only one state and its repetitions mean the robots in the reached cells will stop.

Example 6.1 *For a better understanding of step (i), consider the scenario from Example 4.4 and the LTL specification for the team* $\phi_4 = \Box\Diamond(y_1 \wedge \Diamond y_2)$.

Again, let us assume that the first two robots in the team are the reduced Büchi automaton that are the same as the initial one given in Figure 4.8(d). With $\kappa = 1$, *Algorithm 6.3 would return a single run in* Γ, $\rho = s_1\ s_1$, *with the prefix and suffix* s_1 -

since the initial state is also a final one, the shortest path to infinitely often visit the final state s_1 is to remain in the initial state.

However, for a single robot, the resulted B after running Algorithm 4.4 in line 1 of Algorithm 6.3 is given in Figure 4.9 (one robot cannot visit the disjoint regions y_1 and y_2 at the same time and set H is detailed in the caption of Figure 4.9). In this case, with $\kappa = 1$ we obtain $\rho = s_1\ s_2\ s_3\ s_2\ s_3$ (with suffix $s_2\ s_3$), in accordance with the fact that it is not possible to find a suffix to return in s_1. ■

Step (ii): For enabling the transition from s_j to s_{j+1} in B, $j = 0, \dots, L_\rho - 1$, the following two conditions should hold:

a) The RMPN system has to reach a final marking $\boldsymbol{m}$ that generates any observation from set $\varrho_B(s_j, s_{j+1}) \subseteq 2^Y$;
b) The intermediate RMPN markings should generate only observations in set $\varrho_B(s_j, s_j) \subseteq 2^Y$, such that s_j is not left to other states than s_{j+1}.

In order to impose/verify an observation at a given reachable marking $\boldsymbol{m}$, let us define for each observation $y_i \in Y$ a binary variable x_i such that

$$x_i = \begin{cases} 1, & \text{if } \boldsymbol{v}_i \cdot \boldsymbol{m} > 0 \\ 0, & \text{otherwise.} \end{cases} \tag{6.8}$$

The following two constraints assign the correct value to x_i:

$$\begin{cases} N \cdot x_i & \geq \boldsymbol{v}_i \cdot \boldsymbol{m} \\ x_i & \leq \boldsymbol{v}_i \cdot \boldsymbol{m} \end{cases}, \tag{6.9}$$

where N is a big number. Notice that if $\boldsymbol{v}_i \cdot \boldsymbol{m} > 0$, the first constraint of (6.9) imposes $x_i = 1$, while if $\boldsymbol{v}_i \cdot \boldsymbol{m} = 0$, the second constraint of (6.9) ensures $x_i = 0$.

We now derive formal equivalences in terms of linear inequalities for step (ii-a). For this, consider a generic subset $S \subseteq 2^Y$. Set S (set of subsets of Y) can be seen as a disjunction of conjunctions of propositions from Y; thus, as a Boolean formula φ_S in DNF (Disjunctive Normal Form). Obviously, φ_S is *True* if $\neg\left(\neg\varphi_S\right)$ is *True*. Observe that $\left(\neg\varphi_S\right)$ is the formula corresponding to $2^Y \setminus S$, i.e. $\varphi_{2^Y \setminus S}$, and we want $\neg\varphi_{2^Y \setminus S}$ to be *True*. However, since $\varphi_{2^Y \setminus S}$ is a Boolean formula for the set $2^Y \setminus S$, it is also a DNF and $\neg\varphi_{2^Y \setminus S}$ will be a CNF (Conjunctive Normal Form) [19]. Inspired from results in [147, 148], $\neg\varphi_{2^Y \setminus S}$ can be written as a set of linear constraints using the variables x_i. For this, here we propose Algorithm 6.4, which is also supported by Example 6.2 for an easy to follow case. Informally, once the complement of S is constructed (line 1), we want any element from this complement to be false (not observed). For each element $\bar{s} \in 2^Y \setminus S$, inequality from line 2 is not affected if observing a proposition from $\bar{s}$ (the right-hand term is increased with 1 due x_i, while the left-hand term is increased due to cardinality of $\bar{s}$). However, for each proposition in $\bar{s}$ that is not observed the right-hand term is not increased, thus

Algorithm 6.4: Constraints for the set S

Input: Set $S \subseteq 2^Y$

Output: A set of linear constraints

1 Compute $2^Y \setminus S$ (complement of S);

2 Add constraints $\sum_{y_i \in \overline{s}} x_i - \sum_{y_j \in (Y \setminus \overline{s})} x_j \leq |\overline{s}| - 1,\ \forall \overline{s} \in \left(2^Y \setminus S\right), \overline{s} \neq \emptyset$

3 If $\emptyset \in \left(2^Y \setminus S\right)$, add constraint $\sum_{y_i \in Y} x_i \geq 1$

helping the satisfaction of the inequality. At the same time, the right-hand term is decreased (rewarded) with 1 if a proposition outside set $\overline{s}$ is observed (due to x_j). Overall, if the inequality from line 2 holds, then conjunction from $\varphi_{2^Y \setminus S}$ corresponding to $\overline{s}$ is false, so a disjunction from CNF $\neg\varphi_{2^Y \setminus S}$ becomes true. If $2^Y \setminus S$ includes element $\emptyset$, then at least one proposition from Y should be observed in order to violate this element of the complement set, i.e. inequality from line 3 should hold. Of course, with a slight abuse of notation, if we considered $\overline{s} = \emptyset$ with $|\overline{s}| = 0$ in line 2, we would obtain the same inequality as in line 3.

In summary, for deciding if the current RMPN observation $||\boldsymbol{V} \cdot \boldsymbol{m}||$ belongs to a given set S, we have first to tie the values of binary variables x_i to RMPN observations, as in the inequalities (6.9). Then, these binary variables should also satisfy the inequalities returned by Algorithm 6.4.

Example 6.2 *Let* $Y = \{y_1, y_2, y_3\}$*and* $S = \left\{\{y_1\}, \{y_1, y_2\}, \{y_1, y_3\}, \{y_1, y_2, y_3\}\right\}$. *Satisfaction of S can be seen as the satisfaction of the following Boolean formula in disjunctive normal form:*

$$(y_1 \wedge \neg y_2 \wedge \neg y_3) \vee (y_1 \wedge y_2 \wedge \neg y_3) \vee (y_1 \wedge \neg y_2 \wedge y_3) \vee (y_1 \wedge y_2 \wedge y_3).$$

Negations of unobserved propositions are included in each conjunctive term, because set S contains (due to LTL formalism) all and only feasible observations, and those negations ensure that some propositions are not observed (rather than being either true or false). The set $2^Y \setminus S = \left\{\{y_2\}, \{y_3\}, \{y_2, y_3\}, \{\emptyset\}\right\}$ *where* $\emptyset$ *is the empty observation for the free space:* $\emptyset \equiv \neg y_1 \wedge \neg y_2 \wedge \neg y_3$. *By double negation, the equivalent CNF Boolean formula corresponding to S (negation of* $2^Y \setminus S$*) is*

$$(y_1 \vee \neg y_2 \vee y_3) \wedge (y_1 \vee y_2 \neg y_3) \wedge (y_1 \vee \neg y_2 \vee \neg y_3) \wedge \neg\emptyset.$$

This formula can be converted to a set of linear constraints by Algorithm 6.4:

$$\begin{cases} -x_1 + x_2 - x_3 \leq 1 - 1 = 0 \\ -x_1 - x_2 + x_3 \leq 1 - 1 = 0 \\ -x_1 + x_2 + x_3 \leq 2 - 1 = 1 \\ \ \ x_1 + x_2 + x_3 \geq 1. \end{cases} \tag{6.10}$$

In general, simplification of the formula can be considered. For example, set S in this example corresponds to the formula y_1; indeed, for binary variables, inequalities (6.10) hold only when $x_1 = 1$, while x_2 and x_3 can be either 0 or 1. ■

Cost function: Having the linear constraints for RMPN that ensure a transition in the Büchi automaton, we will develop a Mixed Integer Linear Programming (MILP) formulation. For this, we now need to establish a cost function. The cost function that we propose has three different terms, but the balance between them may result in different solutions.

- The first term considers the minimization of the total number of transition firings. Since σ is the firing vector, minimization of the total number of transition firings can be imposed by minimizing

  $$\alpha \cdot \mathbf{1}^T \cdot \sigma,$$

  where α is a user-chosen weighting value for this term.
- The second term in the cost function minimizes the number of robots that change their position from $\boldsymbol{m}_0$ to $\boldsymbol{m}$. The robots should synchronize when reaching marking $\boldsymbol{m}$ in order to execute the transition in the Büchi automaton, and thus if the number of moving robots is smaller then the synchronization is easier. In order to tackle this aspect, the term $\beta \cdot \parallel \boldsymbol{m}_0 - \boldsymbol{m} \parallel_1$ can be added to the cost function. This L1-norm can be transformed to linear inequalities by adding two vectorial constraints,

  $$\begin{cases} \boldsymbol{m}_0 - \boldsymbol{m} \leq \boldsymbol{w} \\ -\boldsymbol{m}_0 + \boldsymbol{m} \leq \boldsymbol{w} \end{cases},$$

  and adding the following term in the cost function:

  $$\beta \cdot \mathbf{1}^T \cdot \boldsymbol{w}.$$

  Constant β is also used to balance the terms in the cost function.
- The third term deals with collision avoidance. The main idea is to ensure that between $\boldsymbol{m}_0$ and $\boldsymbol{m}$ the number of robots that pass through each cell is as small as possible. If at most one token passes through each RMPN place, and assuming that at $\boldsymbol{m}_0$ this also holds, then there will be no collision. In order to obtain a feasible problem, we will minimize the maximum number of robots that pass through regions instead of imposing a capacity of one in each region. For this, let us define a cumulative vector of markings, $\boldsymbol{m}_t$, that contains non-zero entries in places that were visited along the RMPN trajectory, including the final marking $\boldsymbol{m}$. Due to the RMPN structure, $\boldsymbol{m}_t = \boldsymbol{Pre} \cdot \sigma + \boldsymbol{m}$. Note that $\boldsymbol{m}_t$ is not a reachable marking of RMPN, simply because it can include more tokens than the number of robots. For minimizing the infinity norm of $\boldsymbol{m}_t$, we impose the constraint

$$\boldsymbol{Pre} \cdot \sigma + \boldsymbol{m} \leq \gamma \cdot \mathbf{1},$$

where $\mathbf{1}$ is a vector of length $|P|$ with all elements equal to one, and we add the following term in the cost function:

$$N \cdot \gamma.$$

Note that this term is weighted by the big number N, in accordance with the importance for the collision avoidance aspect. If the solution minimizing the cost function yields $\gamma > 1$, it means that from $\boldsymbol{m}_0$ to $\boldsymbol{m}$ there exists a cell that is crossed by more than one robot and collisions may appear. To cope with this, new intermediate markings can be introduced between $\boldsymbol{m}_0$ and $\boldsymbol{m}$ and the optimization problem can be solved again, or results from Resource Allocation Systems can be used on the computed trajectories [32, 38, 47, 99, 123, 139, 180] (see Section 6.6).

The above aspects allow us to formulate condition (a) of step (ii) as the MILP problem (6.11):

$$\begin{aligned}
& \min \alpha \cdot \mathbf{1}^T \cdot \sigma + \beta \cdot \mathbf{1}^T \cdot \boldsymbol{w} + N \cdot \gamma \\
\text{s.t.} \quad & \boldsymbol{m} = \boldsymbol{m}_0 + \boldsymbol{C} \cdot \sigma \\
& N \cdot x_i \geq \boldsymbol{v}_i \cdot \boldsymbol{m}, \forall y_i \in Y \\
& x_i \leq \boldsymbol{v}_i \cdot \boldsymbol{m}, \forall y_i \in Y \\
& \text{Lin. ineq. in } x_i \text{ given by Algorithm 6.4 for set } \varrho_B(s_j, s_{j+1}) \\
& \boldsymbol{m}_0 - \boldsymbol{m} \leq \boldsymbol{w} \\
& -\boldsymbol{m}_0 + \boldsymbol{m} \leq \boldsymbol{w} \\
& \boldsymbol{Pre} \cdot \sigma + \boldsymbol{m} \leq \gamma \cdot \mathbf{1} \\
& \boldsymbol{m} \in \mathbb{N}_{\geq 0}^{|P|}, \sigma \in \mathbb{N}_{\geq 0}^{|T|}, x_i \in \{0, 1\}, i = 1, \dots, |Y| \\
& \boldsymbol{w} \in \mathbb{R}^{|P|}, \gamma \in \mathbb{R}.
\end{aligned} \tag{6.11}$$

Notice that instead of including constraints on all binary variables x_i, $i = 1, \dots, |Y|$, one can include only those for propositions appearing in $\varrho_B(s_j, s_{j+1})$ (this is not performed in MILP (6.11) for maintaining simpler notations).

MILP (6.11) is solved by using specific optimization routines [102, 150]. The obtained firing vector σ is then projected to individual robot transitions and sequences of visited places, by iteratively firing each enabled transition in the current marking by applying Algorithm 6.1. In rare cases, this projection might not be possible due to the cost function and constraints concerning $\boldsymbol{w}$ and γ, when it is said that σ is *spurious*, a case which will be later handled by Algorithm 6.6. If a solution is obtained, it means that by firing transitions from σ, the RMPN model reaches a final marking $\boldsymbol{m}$ in which the observation enables transition from s_j to s_{j+1} in B.

However, we have to check if the RMPN's intermediate observations did not trigger a transition in B that leaves the current run ρ, i.e. to a state other than s_j or

s_{j+1}. To this goal, we use Algorithm 6.5, whose idea is next explained. Obviously, if no transitions are fired in σ, no further checks are needed (lines 1-2 in Algorithm 6.5) as robots do not move, meaning the transition from s_j to s_{j+1} is already enabled while in $\boldsymbol{m}_0$. Lines 3-4 project σ to individual transitions and sequences of visited places in RMPN by running Algorithm 6.1. Then, on lines 5-9 we find the set of intermediate outputs generated by each robot. While doing this, we assume that the robots individually move (without any intermediate synchronization) and they synchronize only on their last transition to fire, such that they synchronously enter in their final places[2]. That is why we do not include the last output of moving robots (lines 8-9), since all the last outputs of the team definitely enable a desired transition from s_j to s_{j+1} of B. The Cartesian product on line 10 gives the set $Interm_{obs} \subseteq 2^Y$ that includes the possible intermediate observations of RMPN that can be generated while the robots individually move based on σ, while synchronizing only in their last fired transition. The test from line 11 holds if the possible intermediate outputs enable the self-loop from s_j to s_j in B, a case in which we say that the *solution given by σ is applicable*. Otherwise (line 13) state s_j may have been already left to another state than s_{j+1}, before the RMPN reaches the final marking $\boldsymbol{m}$ of MILP (6.11).

If σ given by MILP (6.11) is not applicable, we append more constraints to those from MILP (6.11), for condition (ii-b). Specifically, we consider the cumulative vector of markings $\boldsymbol{m}_t$ defined before to impose restrictions on trajectory observations, without including in $\boldsymbol{m}_t$ the final marking $\boldsymbol{m}$. Thus, we obtain MILP (6.12), which has twice more binary variables than MILP (6.11), the $x_{i(t)}$ variables corresponding to truth values of propositions given by $\boldsymbol{m}_t$:

$$
\begin{aligned}
\min\ \alpha \cdot \mathbf{1}^T \cdot \sigma & + \beta \cdot \mathbf{1}^T \cdot \boldsymbol{w} + N \cdot \gamma \\
\text{s.t.} \qquad \boldsymbol{m} & = \boldsymbol{m}_0 + \boldsymbol{C} \cdot \sigma \\
N \cdot x_i & \geq \boldsymbol{v}_i \cdot \boldsymbol{m}, \forall y_i \in Y \\
x_i & \leq \boldsymbol{v}_i \cdot \boldsymbol{m}, \forall y_i \in Y \\
\text{Lin. ineq. in } x_i \text{ given } & \text{ by Algorithm 6.4 for set } \rho(s_j, s_{j+1}) \\
N \cdot x_{i(t)} & \geq \boldsymbol{v}_i \cdot (\boldsymbol{Pre} \cdot \sigma), \forall y_i \in Y \\
x_{i(t)} & \leq \boldsymbol{v}_i \cdot (\boldsymbol{Pre} \cdot \sigma), \forall y_i \in Y \\
\text{Lin. ineq. in } x_{i(t)} \text{ given } & \text{ by Algorithm 6.4 for set } \rho(s_j, s_j) \\
\boldsymbol{m}_0 - \boldsymbol{m} & \leq \boldsymbol{w} \\
-\boldsymbol{m}_0 + \boldsymbol{m} & \leq \boldsymbol{w} \\
\boldsymbol{Pre} \cdot \sigma + \boldsymbol{m} & \leq \gamma \cdot \mathbf{1} \\
\boldsymbol{m} \in \mathbb{N}_{\geq 0}^{|P|}, \sigma \in \mathbb{N}_{\geq 0}^{|T|}, & \ x_i, x_{i(t)} \in \{0,1\}, i = 1, \ldots, |Y| \\
\boldsymbol{w} \in \mathbb{R}^{|P|}, \gamma \in \mathbb{R}. &
\end{aligned}
\tag{6.12}
$$

2 Such final synchronizations are needed in some cases of LTL formulas, e.g. requiring synchronous visits to some disjoint regions [122].

Algorithm 6.5: Check if σ returned by MILP (6.11) is applicable

Input: σ, RMPN model $\mathcal{Q}$, set $\varrho_B(s_j, s_j)$
Output: Applicability of σ

if $\sigma = \mathbf{0}$ **then**
  σ is applicable;
Run Algorithm 6.1 (project σ to individual robot firing sequences);
Let $seq_i = seq_i[1], seq_i[2], \dots, seq_i[|seq_i|]$ be the sequence of places visited by ith robot, according to its firing sequence;
for *each robot i*, $i = 1, \dots, R$ **do**
  if $|seq_i| = 1$ **then**
    $Obs_i = h(seq_i[1])$;
  else
    $Obs_i = \bigcup_{k=1}^{|seq_i|-1} h(seq_i[k])$;
$Interm_{obs} = Obs_1 \times Obs_2 \times \dots \times Obs_R$;
if $Interm_{obs} \subseteq \varrho_B(s_j, s_j)$ **then**
  σ is applicable;
else
  σ is not applicable;

Informally, MILP (6.12) includes conditions from MILP (6.11) and condition (b) of step (ii). Thus, MILP (6.11) may be computationally faster to solve than MILP (6.12) (due to its fewer variables and constraints), while MILP (6.12) cannot return a non-applicable firing count vector.

The constraints on variables $x_{i(t)}$, together with the cost function and constraints on $\boldsymbol{w}$, γ could imply the execution of some cycles during the trajectories, which is the case when σ is spurious. For spurious σ obtained from MILP (6.11) or (6.12), we can use MILP (6.13), which is similar to MILP (6.5). It includes a number of κ intermediate markings of RMPN ($\boldsymbol{m}_0$, $\boldsymbol{m}_1, \dots, \boldsymbol{m}_{k-1}$) for reaching the final position $\boldsymbol{m} = \boldsymbol{m}_k$. Each robot is constrained to take at most one transition between two successive intermediate markings, and these transitions are synchronized. Moreover, MILP (6.13) replaces the cumulative vector of markings $\boldsymbol{m}_t$ from MILP (6.12) with a sum over intermediate markings reached by RMPN. Thus, whenever MILP (6.13) returns a solution σ, this is an applicable firing count vector:

$$\min \alpha \cdot \mathbf{1}^T \cdot \sum_{i=1}^{k} \sigma_i + \beta \cdot \mathbf{1}^T \cdot \boldsymbol{w} + N \cdot \sum_{i=1}^{k} \gamma_i$$

$$\boldsymbol{m}_i = \boldsymbol{m}_{i-1} + \boldsymbol{C} \cdot \sigma_i, i = 1, \dots, k$$

$$\boldsymbol{m}_{i-1} - \boldsymbol{Pre} \cdot \sigma_i \geq 0, i = 1, \dots, k$$

$$
\begin{aligned}
&N \cdot x_i \geq \boldsymbol{v}_i \cdot \boldsymbol{m}_k, \forall y_i \in Y \\
&x_i \leq \boldsymbol{v}_i \cdot \boldsymbol{m}_k, \forall y_i \in Y \\
&\text{Lin. ineq. in } x_i \text{ given by Algorithm 6.4 for set } \rho(s_j, s_{j+1}) \\
&N \cdot x_{i(t)} \geq \boldsymbol{v}_i \cdot \left(\sum_{i=0}^{k-1} \boldsymbol{m}_i \right), \forall y_i \in Y \\
&x_{i(t)} \leq \boldsymbol{v}_i \cdot \left(\sum_{i=0}^{k-1} \boldsymbol{m}_i \right), \forall y_i \in Y \\
&\text{Lin. ineq. in } x_{i(t)} \text{ given by Algorithm 6.4 for set } \rho(s_j, s_j) \\
&\boldsymbol{m}_0 - \boldsymbol{m}_k \leq \boldsymbol{w} \\
&-\boldsymbol{m}_0 + \boldsymbol{m}_k \leq \boldsymbol{w} \\
&\boldsymbol{m}_i \leq \gamma_i \cdot \boldsymbol{1}, i = 1, \dots, k \\
&\boldsymbol{m}_i \in \mathbb{N}_{\geq 0}^{|P|}, \sigma_i \in \mathbb{N}_{\geq 0}^{|T|}, i = 1, \dots, k\, x_i, x_{i(t)} \in \{0,1\}, i = 1, \dots, |Y| \\
&\boldsymbol{w} \in \mathbb{R}^{|P|}, \gamma_i \in \mathbb{R}, i = 1, \dots, k.
\end{aligned} \tag{6.13}
$$

Note that MILP (6.13) is in general computationally more demanding than MILPs (6.11) and (6.12), due to its additional constraints and unknowns. Note that the value κ is a design parameter. If κ is too small, MILP (6.13) can be unfeasible because from $\boldsymbol{m}_0$ a marking with observations of $\boldsymbol{m}_k$ cannot be reached in so few steps by the RMPN model. One can try to solve MILP (6.13) by starting with $\kappa \geq \boldsymbol{1} \cdot \sigma + R$, where σ is the (unfeasible) solution of MILP (6.12). Note that even for a large κ, MILP (6.13) may fail to provide a solution, simply because the RMPN cannot yield observations that satisfy condition (a) of step (ii) (e.g. one robot cannot be in two disjoint regions/places at the same time/marking).

Step (iii) of our solution just converts the σ vector to sequences of firings for each robot by applying Algorithm 6.1. Then it appends these sequences with previous firing sequences of robots and it imposes that the moving robots (those taking at least one transition) synchronize when taking their last transitions from current σ. In the case where σ is given by MILP (6.13), there will be more (at most κ) synchronizations added for each robot.

Putting together all of the above main steps and the MILPs derived from requirements of step (ii), we obtain the pseudo-code from Algorithm 6.6. Informally, note that cases of spurious solutions from MILPs (6.11) or (6.12), as well as non-applicable solutions of MILP (6.11), are collected by MILP (6.13). However, in all the tests we have performed, MILP (6.13) was not reached, a fact that suggests that the faster optimizations (6.11) and (6.12) might always provide a feasible solution. If Algorithm 6.6 does not return a solution, the problem is deemed unfeasible (the formula may be impossible for the current robotic team and environment or κ should be increased). Note that some runs r of B can be

Algorithm 6.6: Iterative construction of solution

Input: RMPN model Q, Büchi automaton B, κ - number of sequences in B
Output: Solution (firing sequence and synchronizations for each robot)

1 Find set Γ containing κ accepted finite runs of B (using k-shortest path algorithm);
2 **while** $\Gamma \neq \emptyset$ **do**
3   Initialize robot sequences as empty and $\boldsymbol{m}_0$ as initial RMPN marking;
4   Pick shortest run $\rho \in \Gamma$, $\rho = s_0 s_1 \cdots s_{L_\rho}$;
5   **for** $j = 0, 1, \dots, L_\rho - 1$ **do**
6   Formulate and solve MILP (6.11);
7   **if** *σ is spurious* **then**
8   Go to line 19;
9   Run Algorithm 6.5 to establish if σ is applicable;
10   **if** *solution σ is applicable* **then**
11   Update robot sequences;
12   **continue** with next j in for loop;
13   **else**
14   Formulate and solve MILP (6.12);
15   **if** *solution σ is not spurious* **then**
16   Update robot sequences;
17   **continue** with next j in for loop;
18   **else**
19   Formulate and solve MILP (6.13);
20   **if** *solution σ is obtained* **then**
21   Update robot sequences;
22   **continue** with next j in for loop;
23   **else**
24   Current transition of r cannot be ensured;
25   **break** the for loop;
26   **if** *all transitions of ρ were ensured* **then**
27   **return** solution;
28   **else**
29     $\Gamma := \Gamma \setminus \{\rho\}$;

ruled out from Γ without solving any MILP, if some transitions require more than R simultaneous observations of disjoint regions.

6.3.2 Simulation examples

This subsection presents some simulation results on RMTool. Figure 6.3 presents an environment containing seven polygonal and convex regions of interest and five fully actuated point robots initially deployed in the points marked with circles. By partitioning this environment in triangular cells by applying Algorithm 3.2 we obtained 54 triangles. Thus, the PMPN model has 54 places and 158 transitions.

We consider the LTL specification

$$\varphi_1 = \Diamond(y_1 \wedge y_2 \wedge y_3) \;\wedge\; \Box\neg(y_4 \vee y_7) \;\wedge\; (\neg y_5 \,\mathcal{U}\, y_6).$$

The requirement imposes that:

i) there exists a time moment when regions y_1, y_2, y_3 are visited,
ii) regions y_4 and y_7 are always avoided,
iii) region y_6 is eventually reached, without previously entering region y_5.

The resulting Büchi automaton has four states and is sketched in Figure 6.4. The inputs that enable the transitions are the following:

- $\varrho_B(s_1, s_1) = \{\emptyset, \{y_1\}, \{y_2\}, \{y_3\}, \{y_6\}, \{y_1,y_2\}, \{y_1,y_3\}, \{y_1,y_6\}, \{y_2,y_3\}, \{y_2,y_6\}, \{y_3,y_6\}, \{y_1,y_2,y_3\}, \{y_1,y_2,y_6\}, \{y_1,y_3,y_6\}, \{y_2,y_3,y_6\}, \{y_1,y_2,y_3,y_6\}\}$;
- $\varrho_B(s_1, s_2) = \{\{y_6\}, \{y_1,y_6\}, \{y_2,y_6\}, \{y_3,y_6\}, \{y_5,y_6\}, \{y_1,y_2,y_6\}, \{y_1,y_3,y_6\}, \{y_1,y_5,y_6\}, \{y_2,y_3,y_6\}, \{y_2,y_5,y_6\}, \{y_3,y_5,y_6\}, \{y_1,y_2,y_3,y_6\}, \{y_1,y_2,y_5,y_6\}, \{y_1,y_3,y_5,y_6\}, \{y_2,y_3,y_5,y_6\}, \{y_1,y_2,y_3,y_5,y_6\}\}$;
- $\varrho_B(s_1, s_3) = \{\{y_1,y_2,y_3\}, \{y_1,y_2,y_3,y_6\}\}$;
- $\varrho_B(s_1, s_4) = \{\{y_1,y_2,y_3,y_6\}, \{y_1,y_2,y_3,y_5,y_6\}\}$;
- $\varrho_B(s_2, s_2) = \{\emptyset, \{y_1\}, \{y_2\}, \{y_3\}, \{y_5\}, \{y_6\}, \{y_1,y_2\}, \{y_1,y_3\}, \{y_1,y_5\}, \{y_1,y_6\}, \{y_2,y_3\}, \{y_2,y_5\}, \{y_2,y_6\}, \{y_3,y_5\}, \{y_3,y_6\}, \{y_5,y_6\}, \{y_1,y_2,y_3\}, \{y_1,y_2,y_5\}, \{y_1,y_2,y_6\}, \{y_1,y_3,y_5\}, \{y_1,y_3,y_6\}, \{y_1,y_5,y_6\}, \{y_2,y_3,y_5\}, \{y_2,y_3,y_6\}, \{y_2,y_5,y_6\}, \{y_3,y_5,y_6\}, \{y_1,y_2,y_3,y_5\}, \{y_1,y_2,y_3,y_6\}, \{y_1,y_2,y_5,y_6\}, \{y_1,y_3,y_5,y_6\}, \{y_2,y_3,y_5,y_6\}, \{y_1,y_2,y_3,y_5,y_6\}\}$;
- $\varrho_B(s_2, s_4) = \{\{y_1,y_2,y_3\}, \{y_1,y_2,y_3,y_5\}, \{y_1,y_2,y_3,y_6\}, \{y_1,y_2,y_3,y_5,y_6\}\}$;
- $\varrho_B(s_3, s_3) = \{\emptyset, \{y_1\}, \{y_2\}, \{y_3\}, \{y_6\}, \{y_1,y_2\}, \{y_1,y_3\}, \{y_1,y_6\}, \{y_2,y_3\}, \{y_2,y_6\}, \{y_3,y_6\}, \{y_1,y_2,y_3\}, \{y_1,y_2,y_6\}, \{y_1,y_3,y_6\}, \{y_2,y_3,y_6\}, \{y_1,y_2,y_3,y_6\}\}$;
- $\varrho_B(s_3, s_4) = \{\{y_6\}, \{y_1,y_6\}, \{y_2,y_6\}, \{y_3,y_6\}, \{y_5,y_6\}, \{y_1,y_2,y_6\}, \{y_1,y_3,y_6\}, \{y_1,y_5,y_6\}, \{y_2,y_3,y_6\}, \{y_2,y_5,y_6\}, \{y_3,y_5,y_6\}, \{y_1,y_2,y_3,y_6\}, \{y_1,y_2,y_5,y_6\}, \{y_1,y_3,y_5,y_6\}, \{y_2,y_3,y_5,y_6\}, \{y_1,y_2,y_3,y_5,y_6\}\}$;
- $\varrho_B(s_4, s_4) = \{\emptyset, \{y_1\}, \{y_2\}, \{y_3\}, \{y_5\}, \{y_6\}, \{y_1,y_2\}, \{y_1,y_3\}, \{y_1,y_5\}, \{y_1,y_6\}, \{y_2,y_3\}, \{y_2,y_5\}, \{y_2,y_6\}, \{y_3,y_5\}, \{y_3,y_6\}, \{y_5,y_6\}, \{y_1,y_2,y_3\}, \{y_2,y_2,y_5\}, \{y_1,y_2,y_6\}, \{y_1,y_3,y_5\}, \{y_1,y_3,y_6\}, \{y_1,y_5,y_6\}, \{y_2,y_3,y_5\}, \{y_2,y_3,y_6\}, \{y_2,y_5,y_6\}, \{y_3,y_5,y_6\}, \{y_1,y_2,y_3,y_5\}, \{y_1,y_2,y_3,y_6\}, \{y_1,y_2,y_5,y_6\}, \{y_1,y_3,y_5,y_6\}, \{y_2,y_3,y_5,y_6\}, \{y_1,y_2,y_3,y_5,y_6\}\}$.

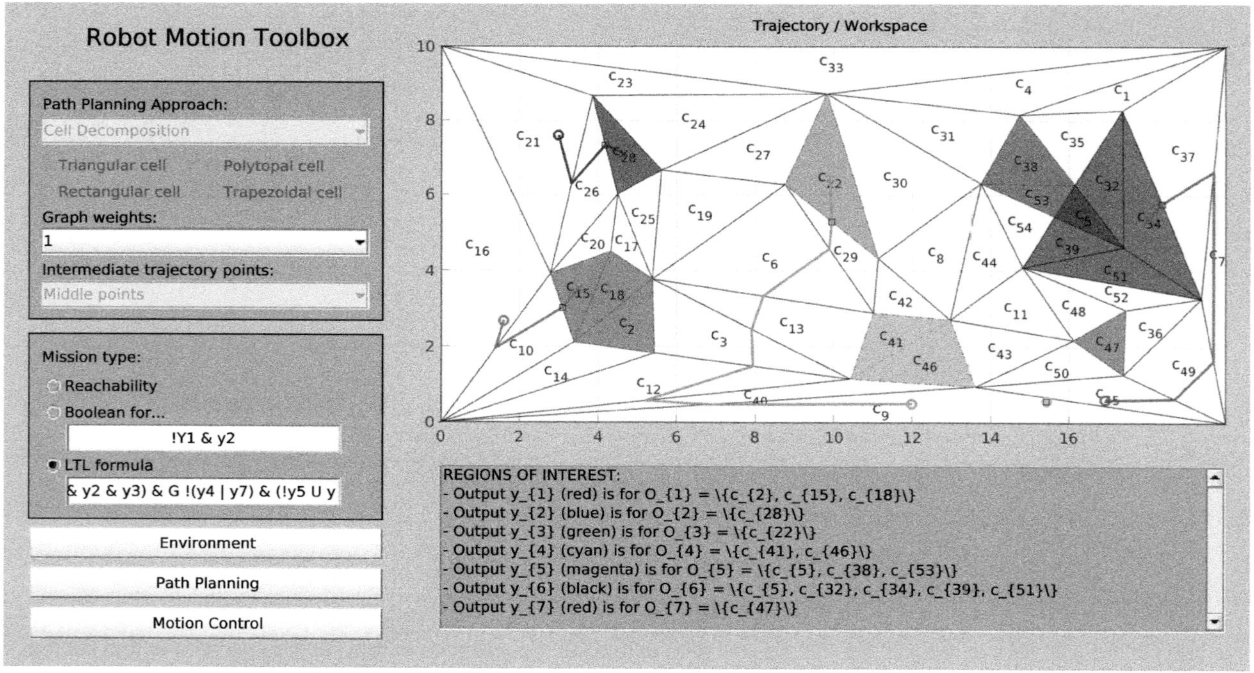

Figure 6.3 Random environment with seven regions of interest and five robots. Trajectories satisfying specification $\varphi_1 = \Diamond(y_1 \wedge y_2 \wedge y_3) \ \wedge \ \Box\neg(y_4 \vee y_7) \ \wedge \ (\neg y_5 \mathcal{U} y_6)$. The robots synchronize when entering their last partition cells, after which they stop in the centroid of the last visited triangle.

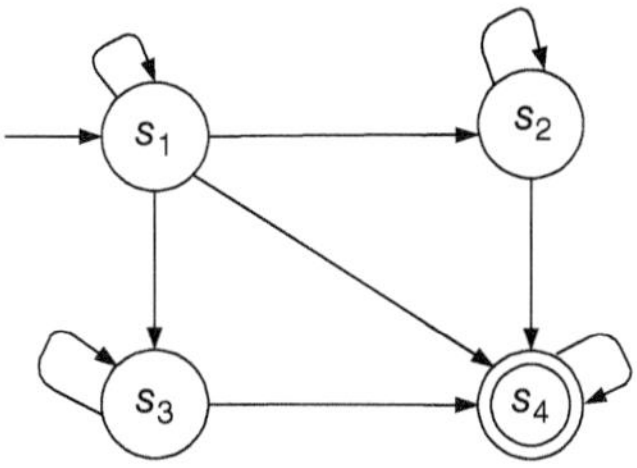

Figure 6.4 Büchi automaton for the LTL specification $\varphi_1 = \Diamond(y_1 \wedge y_2 \wedge y_3) \wedge \Box\neg(y_4 \vee y_7) \wedge (\neg y_5 \; \mathcal{U} \; y_6)$. The inputs that enable transitions in the Büchi automaton are given in Subsection 6.3.2.

Assume $\kappa = 2$ and due to lines 1 of Algorithm 6.6, $\Gamma = \{\{s_1 s_4 s_4\}, \{s_1 s_2 s_4 s_4\}$. Line 4 of the algorithm choose, the shortest run, which in this case contains the initial and the final states, i.e. $\rho = s_1 s_4 s_4$ with prefix $s_1 s_4$ and suffix s_4. From the B detailed structure, we obtain $\varrho_B(s_1, s_4) = \big\{\{y_1, y_2, y_3, y_6\}, \{y_1, y_2, y_3, y_5, y_6\}\big\}$; indeed, the formula is satisfied if any of these two observations is generated by RMPN, with no other region visited until then.

For the first transition of ρ (i.e. $s_1 \to s_4$), MILP (6.11) is constructed, having 274 unknowns ($\boldsymbol{m}$, σ, $x_1, \ldots, x_7$, $\boldsymbol{w}$, γ), while constraints include 54 equalities (the state equation of RMPN) and 302 inequalities given by Eq. (6.9), Algorithm 6.4, and constraints for $\boldsymbol{w}$ and γ. By choosing $\alpha = \beta = 1$ and $N = 1000$, a solution is obtained in about 0.04 seconds, and the returned σ is not applicable by Algorithm 6.5. MILP (6.11) returns the following trajectories for the robots:

- $r_1 : c_{16} c_{10} c_{15}$;
- $r_2 : c_{21} c_{26} c_{28}$;
- $r_3 : c_9 c_{40} c_{46} c_{41} c_{42} c_{29} c_{22}$;
- $r_4 : c_9$;
- $r_5 : c_{45} c_{49} c_{36} c_{52} c_{51}$.

Notice that the final regions reached by the robots activate the transition in the Büchi automaton. However, region y_4 was generated along robot trajectories (since such trajectory restrictions are not captured by MILP (6.11)). In particular, robot r_3 crosses regions c_{46} and c_{41} corresponding to the output y_4. Advancing in Algorithm 6.6, MILP (6.12) runs at line 14. MILP (6.12) has 281 unknowns, 54 equality constraints, and 428 inequality constraints and gets a solution in approximately 0.14 seconds for the first transition in ρ. In particular, the trajectories for robots r_1, r_2, and r_4 are the same as before. However, now

- $r_3 : c_9 c_{40} c_{12} c_3 c_{13} c_6 c_{29} c_{22}$;
- $r_5 : c_{45} c_{49} c_7 c_{37} c_{34}$.

Since these new trajectories are feasible, Algorithm 6.6 continues with the second transition in ρ (i.e. $s_4 \to s_4$). MILP (6.11) returns this time a feasible solution in less than one second and the robots are not changing the regions reached before.

The corresponding robot trajectories are represented in Figure 6.3, by simply connecting the middle points of the line segments shared by successive cells [35]. Recall that all the robots simultaneously enter their last visited cell, thus guaranteeing that the first element of $\varrho_B(s_1, s_4)$ is obtained and until then the RMPN output is $\emptyset$.

Let us consider now the following specification:

$$\varphi_2 = \Diamond(y_1 \wedge y_2 \wedge y_3) \ \wedge \ \Box\neg(y_4 \vee y_7) \ \wedge \ \neg(y_5 \vee y_6) \ \mathcal{U} \ (y_5 \wedge y_6).$$

Compared to φ_1, this requirement changes only the last specification to (iii) regions labeled y_5 and y_6 are simultaneously entered. Automaton B has also four states, and its shortest run has $\varrho_B(s_1, s_4) = \left\{\{y_1, y_2, y_3, y_5, y_6\}\right\}$ and $\varrho_B(s_4, s_4) = \emptyset$, i.e. robots stop in the final regions. MILP (6.11) is solved in 0.3 seconds, but the returned σ is not applicable, as with Algorithm 6.5. Intuitively, this is because three robots would be driven towards y_1, y_2, y_3, respectively, another robot would be driven towards place p_5 - the intersection of regions y_5 and y_6 - and one robot would not move. However, place p_5 cannot be directly reached from the free space and thus the corresponding robot would yield observation y_6 before $(y_5 \wedge y_6)$, thus violating the requirement (iii) of φ_2. Thus, Algorithm 6.6 formulates and solves MILP (6.12). A solution is obtained in less than 1 second, which gives the trajectories represented in Figure 6.5, where the robots synchronize in positions marked with "□", i.e. when entering their last visited places.

For this last example, if we removed the last two terms from the cost function MILP (6.12) (similar to making $\beta = N = 0$), we would obtain the trajectories from Figure 6.6, where in addition two robots could collide in p_{21}.

Note that a transition system approach based on automaton products [54] (presented in Section 5.6) cannot be used in the above scenarios. This is because each robot would be described by a transition system with 54 states, and the possible movements of the whole team would require a transition system with 54^5 states. This number would then be multiplied by the size of automaton B, and hence the state explosion problem of handling an automaton with more than 10^9 states. Thus, the approach presented in this Section and based on RMPN models and MILP formulations brings a real benefit, by providing solutions in a small enough computation time.

6.4 A sequencing problem

6.4.1 Problem statement

In this section we assume a partitioned environment in which some of the regions could contain some resources. These resources should be collected by a team of robots in a given order. Assume, for example, that the environment in Figure 6.7

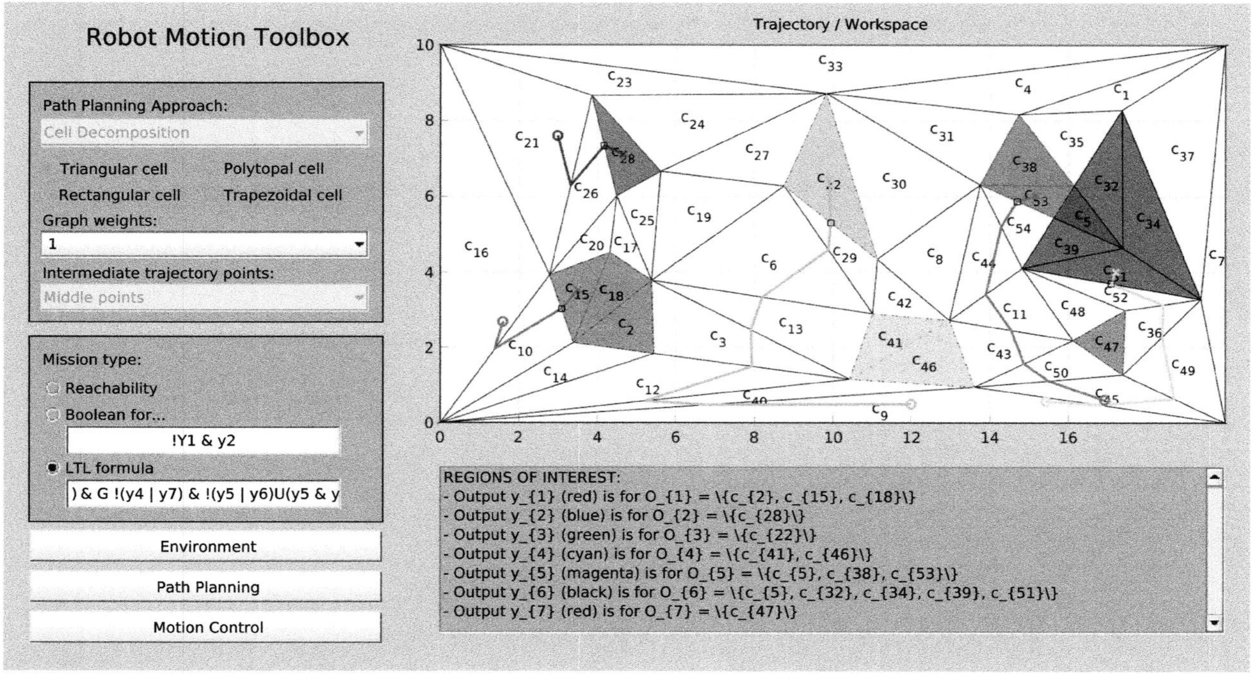

Figure 6.5 Trajectories computed for the LTL specification $\varphi_2 = \Diamond(y_1 \wedge y_2 \wedge y_3) \;\wedge\; \Box\neg(y_4 \vee y_7) \;\wedge\; \neg(y_5 \vee y_6)\;\mathcal{U}\;(y_5 \wedge y_6$. The robots synchronized on the last movement before entering their last region.

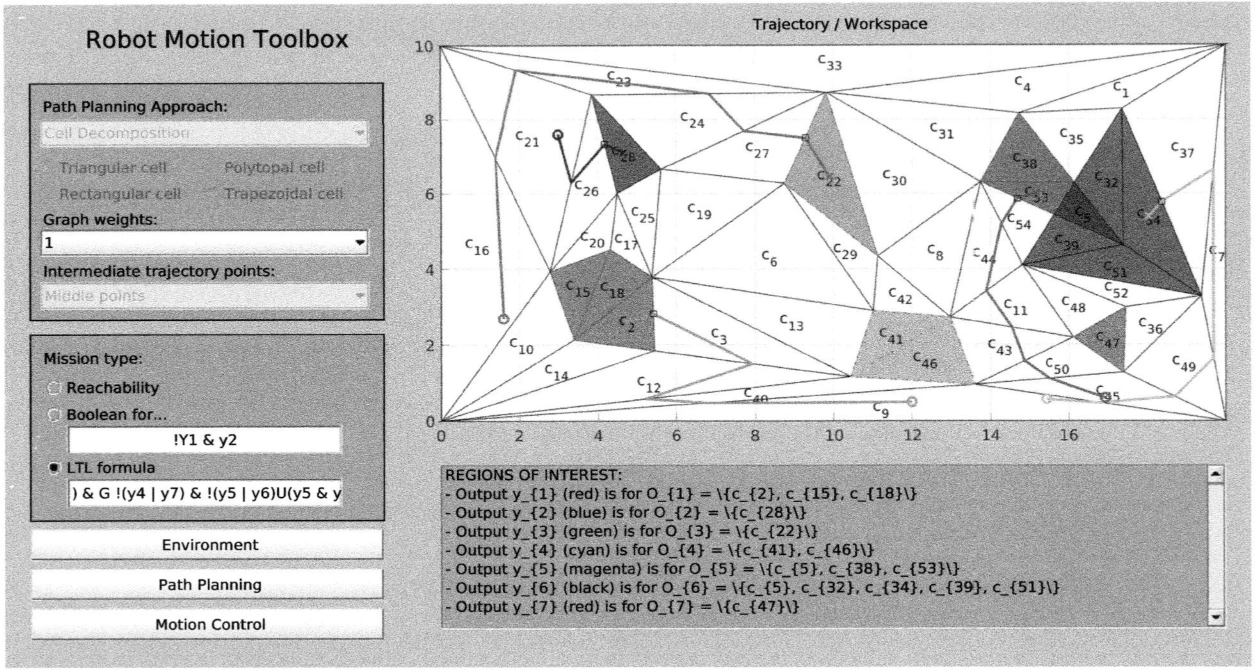

Figure 6.6 Trajectories computed for the LTL specification $\varphi_2 = \Diamond(y_1 \wedge y_2 \wedge y_3) \ \wedge \ \Box\neg(y_4 \vee y_7) \ \wedge \ \neg(y_5 \vee y_6) \ \mathcal{U} \ (y_5 \wedge y_6$. Other trajectories satisfying specification φ_2, obtained without considering the maximum number of robots in RMPN places. Different from trajectories in Figure 6.5, the red and blue robots (left side) can collide in c_{21}.

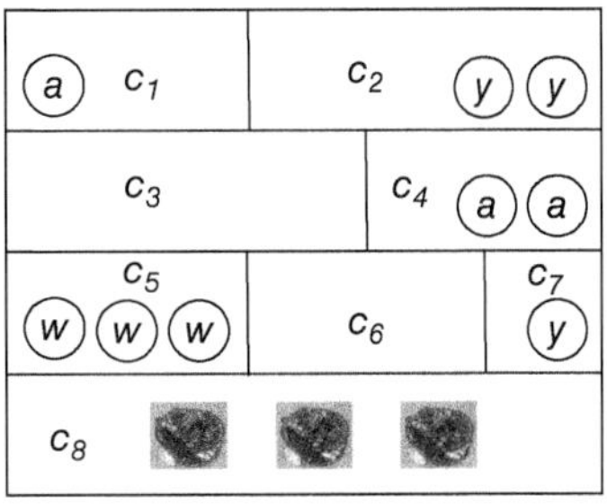

Figure 6.7 A simple environment to illustrate the sequencing problem. An environment composed of eight regions, three types of resources, and three robots initially placed in c_8.

is partitioned into eight regions $C = \{c_1, c_2, \dots, c_8\}$ with three identical robots initially placed in c_8. There are three types of resources in terms of some letters, in this case resources of type 'w', resources of type 'a', and resources of type 'y'. For example, region c_5 contains three resources of type 'w' while region c_4 contains two resources of type 'a'. To model this, we can define three regions of interest $O = \{O_1, O_2, O_2\}$ where $O_1 = \{c_5\}$ (for type 'w'), $O_2 = \{c_1, c_4\}$ (for type 'a'), and $O_3 = \{c_2, c_7\}$ (for type 'y'), respectively. Based also on the adjacency relations between regions, a graph $G = \langle C, E, O\rangle$ as in Definition 4.1 can be defined.

By applying Algorithm 4.2 with G and the adjacency matrix as input parameters, the RMPN $Q = \langle \mathcal{N}, \boldsymbol{m}_0, H, h\rangle$ is obtained, as given in Definition 4.4 and in Figure 6.8. We assume that the regions of interest are *disjoint*, i.e. $O_i \cap O_j = \emptyset$, $\forall i \neq j$. Let $P_k = \{p \in P | h(p) = y_k\}$ be the set of places (regions) corresponding to the region of interest O_k, and hence with observation y_k.

Since each region of interest $O_i \in O$ could contain some resources (e.g. pieces or samples) of type i, this can be formally given by a set of functions η_i, $i = 1, \dots, |O|$, such that $\eta_i : P_i \to \mathbb{N}$. Function η_i is giving the number of resources of type i available in places belonging to P_i. In particular, $\eta_i(p)$ with $p \in P_i$ is the number of resources of type i in place p. Obviously, the total number of resources of type i is given by

$$\sum_{p \in P_i} \eta_i(p).$$

For the RMPN in Figure 6.8, $P_1 = \{p_5\}$ (observation y_1 could be active only in p_5), $P_2 = \{p_1, p_4\}$, while $P_3 = \{p_2, p_7\}$. Since in p_5 there are three resources of type 1, then $\eta_1(p_5) = 3$ while $\eta_2(p_5) = \eta_3(p_5) = 0$ because there is no resource of type 2 or of type 3 in place p_5.

Assuming a team $R = \{r_1, r_2, \dots, r_{|R|}\}$ of identical mobile robots evolving in this environment, the task is to collect and assemble resources from places P_i under these restrictions:

- A robot has to first collect a resource of type 1, then transport it to a place from P_2 and assemble the pieces of types 1 and 2, transport the assembled resources to a region of type 3, and so on until type $|O|$.

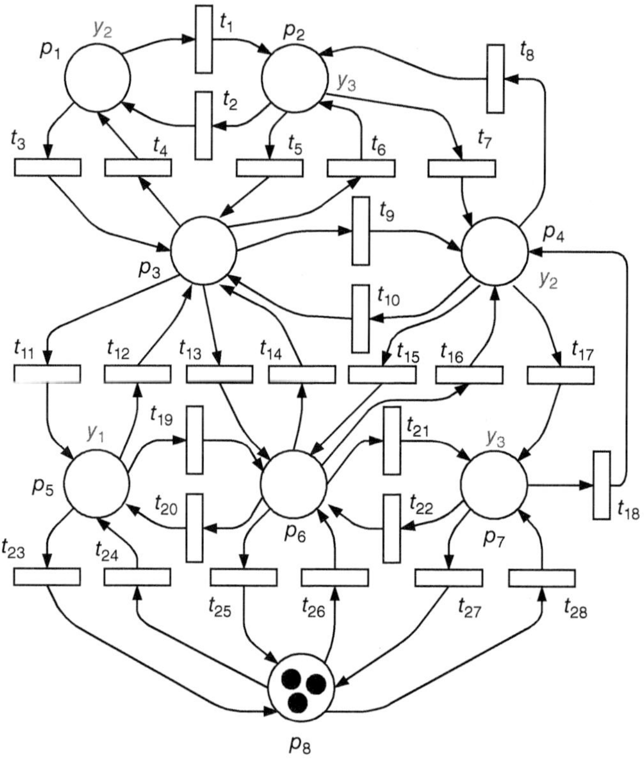

Figure 6.8 RMPN for the team in the environment illustrated in Figure 6.7. The RMPN consists of 8 places, 28 transitions, and three tokens in p_8 corresponding to the three robots. The output function is $h(p_1) = y_2$, $h(p_2) = y_3$, $h(p_4) = y_2$, $h(p_5) = y_1$, and $h(p_7) = y_3$ corresponding to the regions of interest in Figure 6.7.

- Resources of different types are thus assembled (collected) in the specific order $1, 2, \dots, |O|$. The result is left in the last visited region (of type $|O|$) and then the robot can iterate the previous step.
- Each robot can carry at most one intermediate assembly, i.e. it cannot collect more resources of the same type, nor can it collect a resource of type $i + 2$ immediately after a resource of type i. However, a robot carrying a resource of type i is not forbidden to visit a place from set P_{i+2}; it can cross such a region without collecting the resource of type $i + 2$ at that moment.
- A robot is not allowed to drop off an unfinished assembly, i.e. once it picks a resource of type 1, it should transport and assemble it with resources of types $2, \dots, |O|$.
- Two robots cannot exchange the collected resources between them.

Note that the specific order of assembling resources is not restrictive, since any other desired ordering is simply accomplished by relabeling sets $P_1, P_2, \dots, P_{|O|}$. For the sake of being able to finish all possible assemblies, we assume equal numbers of resources of different types, i.e.

$$\sum_{p\in P_i} \eta_i(p) = \sum_{p\in P_j} \eta_j(p), \forall i,j \in \{1,2,\dots,|O|\},\ i \neq j.$$

Furthermore, as mentioned before, we assume that regions of different types do not intersect, i.e.

$$P_i \cap P_j = \emptyset, \forall i,j \in \{1,\dots,|O|\},\ i \neq j,$$

and that no robot is initially placed in any region from P_1.

In the RMPN in Figure 6.8, the robots should collect first a "w" (resource of type 1), go to *assemble* with an "a" (resource of type 2), and finally go to collect a "y" (resource of type 3) to create the word "way".

The outcome of the solution provided in this section will consist in individual plans for each robot such that all resources are eventually assembled. Due to the finite-state representation, we do not explicitly capture robot collisions, since doing this at the abstraction level would be too conservative. In a real scenario, one can assume local collision-avoidance rules (e.g. based on sensorial readings) or results as presented in Section 6.6.

The usefulness of the above problem can range from manufacturing scenarios, where different parts should be collected and assembled in a specific order, to more general missions that involve tasks such as search-and-rescue or collect and drop off medical aids in specific regions.

6.4.2 Solution

We propose an iterative solution for the previous problem. This solution plans the robots movements based on the RMPN model, by iterating a reachability solution.

Reachability of desired states. A linear programming problem (LPP) can be used to reach some desired regions (places). Let us assume that the robots are initially located in regions given by $\boldsymbol{m}_0$ and that they should reach regions where each robot can collect a resource of type i. Observe that if the number of robots $|R| \leq \sum_{p\in P_i} \eta_i(p)$ then all robots will move (there are more resources than robots), while otherwise only a subset of robots should move.

Because the RMPN model given in Definition 4.4 is a state machine, we show that this reachability problem can be translated to an LPP formulation instead of a MILP. Reaching the set of desired places P_i is the same as reaching a desired marking $\boldsymbol{m}$ that satisfies:

i) $\boldsymbol{m} = \boldsymbol{m}_0 + \boldsymbol{C} \cdot \sigma$ - final marking is the solution of the state equation.
ii) The set of robots that reach places from P_i is equal to $\tilde{r}$, where

$$\tilde{r} = min\left(|R|, \sum_{p \in P_i} \eta_i(p)\right),$$

where min denotes the minimum of two numbers. Using the characteristic vector of output y_i defined in Section 4.4, this can be imposed by using the following constraint:

$$\boldsymbol{v}_i \cdot \boldsymbol{m} = \tilde{r}.$$

This equality imposes $\tilde{r}$ robots that should be located at $\boldsymbol{m}$ in places where observation y_i is active.
iii) $\boldsymbol{m}[p] \leq \eta_i(p), \forall p \in P_i$ - there exist enough resources of type i in each region from P_i reached by robots. The number of robots outside P_i is unconstrained and the previous inequalities can be written in the following vector form:

$$\boldsymbol{m} \leq \eta_i,$$

where η_i is a vector such that

$$\eta_i[p] = \begin{cases} \eta_i(p), & \forall p \in P_i \\ \infty, & \forall p \in P \setminus P_i. \end{cases} \tag{6.14}$$

There is a whole set of desired markings corresponding to set P_i and we find one that can be reached by firing a minimum number of transitions in RMPN system Q. Denoting by σ the firing vector that yields the desired marking to be reached, the following LPP results:

$$\begin{aligned} &\min \boldsymbol{1}^T \cdot \sigma \\ \text{s.t. } &\boldsymbol{m} - \boldsymbol{C} \cdot \sigma = \boldsymbol{m}_0 \\ &\boldsymbol{v}_i \cdot \boldsymbol{m} = \tilde{r} \\ &\boldsymbol{m} \leq \eta_i \\ &\boldsymbol{m}, \sigma \geq \boldsymbol{0}. \end{aligned} \tag{6.15}$$

Proposition 6.1 Let Q be an RMPN system as in Definition 4.4, modeling the environment with $|R|$ robots. Assume that the robots should reach some places from P_i to pick up resources. The constraints of LPP (6.15) define an integer convex polytope[3]. Additionally, the LPP (6.15) solved with a Simplex algorithm returns an integer optimal solution for reaching desired regions.

Proof: The above statement can be proved using two basic results.

3 An integer polytope has all vertices formed by integer numbers.

- Let us consider the polyhedron

$$Q(\boldsymbol{A}, \boldsymbol{b}, \boldsymbol{b}', \boldsymbol{c}, \boldsymbol{c}') = \\ = \{\boldsymbol{x} \mid \boldsymbol{b} \leq \boldsymbol{A} \cdot \boldsymbol{x} \leq \boldsymbol{b}' \text{and } \boldsymbol{c} \leq \boldsymbol{x} \leq \boldsymbol{c}'\},$$

  where $\boldsymbol{A}$ is a matrix of integer numbers and the entries of vectors $\boldsymbol{b}$, $\boldsymbol{b}'$, $\boldsymbol{c}$, and $\boldsymbol{c}'$are either integer numbers or $\pm\infty$. Theorem 2 in [96] states that $Q(\boldsymbol{A}, \boldsymbol{b}, \boldsymbol{b}', \boldsymbol{c}, \boldsymbol{c}')$ is an integer polyhedron if and only if $\boldsymbol{A}$ is totally unimodular.
- Let $\boldsymbol{A}$ be a matrix whose rows can be partitioned into two disjoint sets $\boldsymbol{B}$ and $\boldsymbol{D}$. According to [93], the following four conditions are sufficient for $\boldsymbol{A}$ to be totally unimodular:
  i) Every column of $\boldsymbol{A}$ contains at most two non-zero entries;
  ii) Every entry in $\boldsymbol{A}$ is 0, +1, or −1;
  iii) If two non-zero entries in a column of $\boldsymbol{A}$ have the same sign, then the row of one is in $\boldsymbol{B}$ and the other in $\boldsymbol{D}$;
  iv) If two non-zero entries in a column of $\boldsymbol{A}$ have opposite signs, then the rows of both are in $\boldsymbol{B}$, or both in $\boldsymbol{D}$.

Let us consider the polytope defined by the constraints of LPP (6.15). The set of constraints can be rewritten as:

$$\begin{cases} \begin{bmatrix} \boldsymbol{I} & -\boldsymbol{C} \\ \boldsymbol{v}_i & \boldsymbol{0} \end{bmatrix} \cdot \begin{bmatrix} \boldsymbol{m} \\ \sigma \end{bmatrix} = \begin{bmatrix} \boldsymbol{m}_0 \\ \tilde{r} \end{bmatrix} \\ \boldsymbol{0} \leq \begin{bmatrix} \boldsymbol{m} \\ \sigma \end{bmatrix} \leq \begin{bmatrix} \eta_i \\ \infty \end{bmatrix} \end{cases} \tag{6.16}$$

By considering $\boldsymbol{A} = \begin{bmatrix} \boldsymbol{I} & -\boldsymbol{C} \\ \boldsymbol{v}_i & \boldsymbol{0} \end{bmatrix}$, $\boldsymbol{b} = \boldsymbol{b}' = \begin{bmatrix} \boldsymbol{m}_0 \\ \tilde{r} \end{bmatrix}$, $\boldsymbol{c} = \boldsymbol{0}$, $\boldsymbol{c}' = \begin{bmatrix} \eta_i \\ \infty \end{bmatrix}$ and unknowns (variables) $\boldsymbol{x} = \begin{bmatrix} \boldsymbol{m} \\ \sigma \end{bmatrix}$, constraints from (6.16) define a polyhedron as $Q(\boldsymbol{A}, \boldsymbol{b}, \boldsymbol{b}', \boldsymbol{c}, \boldsymbol{c}')$ from above. Because the RMPN system $\mathcal{Q}$ is a strongly connected state machine, each column of matrix $\boldsymbol{C}$ has two non-zero elements, one is −1 and the other is +1. Thus, matrix $\boldsymbol{A}$ satisfies conditions (i) and (ii), and its rows can be partitioned in $\boldsymbol{B} = [\boldsymbol{I} \ -\boldsymbol{C}]$ and $\boldsymbol{D} = [\boldsymbol{v}_i \ \boldsymbol{0}]$ to satisfy condition (iv). Therefore, $\boldsymbol{A}$ is a totally unimodular matrix and thus the polytope defined by the constraints in LPP (6.15) is an integer. If LPP (6.15) is solved by the Simplex algorithm, the solution is a vertex of this polytope [42], and thus the resulting $\boldsymbol{m}$ and σ have positive integer elements. Since RMPN $\mathcal{Q}$ is a live state machine, marking $\boldsymbol{m}$ returned by LPP (6.15) is reachable by firing transitions according to σ. The firing sequences can be obtained by running Algorithm 6.1. ■

Iterative solution. An algorithmic solution for the full problem can be obtained by iterating (for each resource type) the previous procedure that sends the robots to collect and assemble the next needed pieces. The method stops when there are no more uncollected resources in the environment.

Algorithm 6.7 gives a pseudo-code description of the iterative solution. The currently targeted type of resources (i) is imposed on the loop from line 2. On lines 3-5 the robot movements are planned such that at most $\eta_i(p)$ robots are sent in each region $p \in P_i$. To accomplish this, LPP (6.15) is used, with $\boldsymbol{m}_0$ from line 3 giving the current RMPN marking (robot positions) and P_i, η_i indicating the set of desired markings to be reached. The cost function from LPP (6.15) implies that the maximum possible number of robots will move, i.e. all robots if there are more than $|R|$ available resources of type i and $\left(\sum_{p \in P_i} \eta_i(p)\right)$ robots otherwise.

After collecting a type of resources (lines 6-8), the number of remaining pieces of the same type is updated (line 9). Also, the initial marking of the RMPN system is updated such that the next planning solution (towards set P_{i+1}) accounts for the current positions of robots that have just collected resources of type i (lines 10 and 11).

If there are robots that are not sent for collecting pieces, these robots are removed from further plans (line 11). This is because such a situation occurs only when there were less uncollected resources of the current type i than the number of robots. Note that the cost function from LPP (6.15) and the assumption that sets P_i, P_j are disjoint, $\forall i, j \in \{1, \ldots, |O|\}$, $i \neq j$, and guarantee that the robots that do not collect pieces remain in their current position (i.e. the corresponding tokens from RMPN model do not move). Thus, any robot outside places in P_i is not further needed (line 11) and will remain stopped.

The current set P_i is updated (decreased) on line 12 such that it points only to regions that still have uncollected pieces of type i. The stopping condition from line 1 means that there are no more unassembled pieces, under the assumption of equal numbers of each resource type. Therefore, Algorithm 6.7 finishes after a finite number of steps.

Due to the iterative optimizations that plan multiple robots at each step, a solution from Algorithm 6.7 does not guarantee optimality from the point of view of the total number of transitions fired in model Q in order to assemble all available resources.

We mention that Algorithm 6.7 can be adjusted such that different numbers of samples of the same type are accommodated. To accomplish this, one should separately execute loops starting on lines 1 and 2 for each non-empty set P_i, $i = 1, \ldots, |O|$. Additionally, the robots that are not allocated in an iteration should not be removed, but stored in a set that can be useful for subsequent resource types.

Remark 6.3 Each iteration of Algorithm 6.7 yields a moving plan for the robotic team until resources of the next type are picked up. Note that in a real implementation, there is no need to wait until all robots collect the aimed resources and

Algorithm 6.7: Sequencing problem on the RMPN system Q

while $|P_1| > 0$ **do**
  for $i = 1, 2, \ldots, |O|$ **do**
    Solve LPP (6.15) for $\boldsymbol{m}_0$, P_i, $\boldsymbol{\eta}_i$
    Solution returned by LPP (6.15): $\boldsymbol{m}$ (final marking), $\boldsymbol{\sigma}$ (firing vector)
    Move robots according to firing vector $\boldsymbol{\sigma}$
    for $j = 1, 2, \ldots, |R|$ **do**
      if robot r_j reached a place $p \in P_i$ **then**
        r_j collects (and assembles) a resource of type i
        $\eta_i(p) := \eta_i(p) - 1$
    $\boldsymbol{m}_0 := \boldsymbol{m}$
    $\boldsymbol{m}_0[p] := 0, \forall p \in P \setminus P_i$
    $P_i := P_i \setminus \{p \in P_i \mid \eta_i(p) = 0\}$

then proceed to the next iteration. Instead, the next iterations can be solved during robotic movements and further plans can be stored and applied when necessary. Also, due to the problem requirements, there is no need to synchronize the robots when a type of resource is collected. Thus, each robot can individually follow its movement plan yielded by all iterations of Algorithm 6.7. If necessary, one can employ local rules for avoiding collisions among team members.

The computational burden of our approach is given by iterating LPPs guided by Algorithm 6.7. Even if is necessary to use Simplex-based algorithms, it is well known that usually such algorithms perform fast. The number of iterations depends on the number of tasks and samples and on the team size.

Simulation example. Let us consider again the RMPN model Q from Figure 6.8. There are eight regions in the environment (represented by places $p_1, \ldots, p_8$), three types of resources (with $P_1 = \{p_5\}$, $P_2 = \{p_1, p_4\}$, $P_3 = \{p_2, p_7\}$), and three resources of each type ($\eta_1(p_5) = 3$, $\eta_2(p_1) = 1$, $\eta_2(p_4) = 2$, $\eta_3(p_2) = 2$, $\eta_3(p_7) = 1$). The team is formed by three robots, initially located in p_8. Transitions $t_1, \ldots, t_{28}$ model the robot movement capabilities between environment regions.

The first iteration of Algorithm 6.7 plans the robots such that they reach the marking with $\boldsymbol{m}[p_5] = 3$ by firing transition t_{24} three times. After collecting three resources of type 1, in the second iteration r_1 is sent to p_1 (via transitions t_{12} and t_4), while r_2 and r_3 are sent to p_4 (via transitions t_{19} and t_{16}), where each robot assembles the collected piece of type 1 with a resource of type 2. The third iteration sends r_1 and r_2 to p_2 while r_3 is sent to p_7, and when they reach these places each of them completes an assembly of pieces of types 1, 2, 3. Appending the robot plans given by iterations of Algorithm 6.7, the individual places and transitions followed by each robot are as follows:

$$\begin{aligned} r_1 &: p_8 \xrightarrow{t_{24}} p_5 \xrightarrow{t_{12}t_4} p_1 \xrightarrow{t_1} p_2, \\ r_2 &: p_8 \xrightarrow{t_{24}} p_5 \xrightarrow{t_{19}t_{16}} p_4 \xrightarrow{t_8} p_2, \\ r_3 &: p_8 \xrightarrow{t_{24}} p_5 \xrightarrow{t_{19}t_{16}} p_4 \xrightarrow{t_{17}} p_7. \end{aligned} \tag{6.17}$$

6.5 Task gathering problem

6.5.1 Problem formulation

Consider an RMPN model as in Definition 4.4 that corresponds to a team of $|R|$ identical robots evolving in a given environment. We assume that in this environment there are n tasks to be accomplished (served), as follows. Each task has a fixed and known location and for this to be accomplished it is necessary to have several robots simultaneously placed in the task location. The location of tasks is given by the function $g : \{1, \dots, n\} \to P$ and the needed number of robots for serving tasks is given by the function $r : \{1, \dots, n\} \to \{1, \dots, |R|\}$. Thus, task i is located in place $g(i)$ and is accomplished only when at least $r(i)$ robots are located in $g(i)$.

As an example of a possible manufacturing scenario, the tasks can correspond to some pieces to be pressed or lifted and, depending on the hardness or weight of each piece, more robots are needed to collaborate in fulfilling the task.

Example 6.3 *Let us consider the same environment as in Figure 6.7, also with three robots but now, instead of letters, let us assume that in some regions some boxes exist with tokens inside. The new environment is given in Figure 6.9. The RMPN for this environment is the same with the RMPN in Figure 6.8 being the same partition. In this case, the tasks ($n = 4$) are defined as follows:*

- *task 1:* $g(1) = \{p_1\}$ *and* $r(1) = 2$;
- *task 2:* $g(2) = \{p_2\}$ *and* $r(2) = 3$;
- *task 3:* $g(3) = \{p_4\}$ *and* $r(3) = 2$;
- *task 4:* $g(4) = \{p_5\}$; *and* $r(4) = 1$.

Problem to solve. Given the model Q and the maps g and r, find optimal sequences of transitions to be taken by the robots such that all n tasks are accomplished. ■

We consider that the optimality is formulated with respect to the total number of transitions to that should be taken by the robots until all tasks are served. If one adds a weight (cost) on each transition (e.g. corresponding to the energy required by a robot to move between the two linked locations), this optimality correspond, to the minimum amount of energy spent for moving the robots.

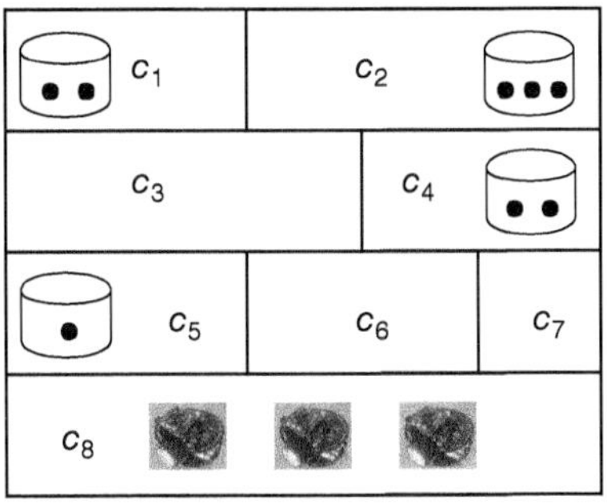

Figure 6.9 A simple environment to illustrate the task gathering problem. An environment composed of eight regions, four tasks ($n = 4$) and three robots initially placed in c_8. The number of robots necessary to fulfill the tasks is equal to the number of tokens in the cylinders, e.g. 2 robots for the task in c_1 and 1 for the task in c_5.

The solution presented in the following subsection will consist individual sequences of transitions to be followed by each robot and in some waiting strategies that are needed for enforcing the required number of robots to arrive at a task location before serving it and moving further. We assume that inter-robot collisions are not of interest at the high level of planning the task executions on the RMPN model, and if necessary they can be avoided by using local rules while robots move. Once each robot receives its movement strategy, the approach becomes decentralized and the waiting modes for robots can be accomplished only by communication inside a single location (task location).

6.5.2 Solution

The solution is based on formulating the problem requirements as a set of linear inequalities. Together with the state equation of the RMPN system (that corresponds to the team capabilities), we obtain a set of constraints for an optimization problem. By adding a cost function for optimizing the number of fired transitions, we will obtain a Mixed Integer Linear Programming (MILP) formulation, whose solution can be used for constructing movement strategies for robots. The proper order of serving the tasks results from the MILP problem, such that all tasks are accomplished by firing a minimum number of transitions in the RMPN system. We mention that if one desires to serve the tasks in a specific order, then the problem becomes a lot simpler, being similar to the problem presented in Section 6.4.

To solve this problem, we first note that during the RMPN evolution we should follow at most n important intermediate markings (like checkpoints), each such marking corresponding to placing more robots in a needed location for solving a task. Of course, it may happen that two or more tasks are simultaneously served by subgroups of robots, or that some tasks can be served in the initial marking, which is the reason we referred to "at most" n important markings. We denote these markings by $\boldsymbol{m}_j, j = 1, \dots, n$, and the firing vectors that lead from one marking to the next by $\sigma_j, j = 1, \dots, n$. That is, by starting from initial marking $\boldsymbol{m}_0$ and firing transitions from σ_1 the RMPN reaches marking $\boldsymbol{m}_1$, and so on. Observe that we do not know yet which task(s) is(are) served at marking $\boldsymbol{m}_1$, because the order of serving tasks will result after solving the MILP problem. Based on the state equation of the RMPN, we have until now the following constraints:

$$\boldsymbol{m}_j = \boldsymbol{m}_{j-1} + \boldsymbol{C} \cdot \sigma_j, \quad j = 1, \dots, n. \tag{6.18}$$

Let us define a set of $n \cdot (n+1)$ binary variables $x_{i,j}$ with the role of witnessing if a given task can be served at a given marking, as follows:

$$x_{i,j} = \begin{cases} 1, & \text{if task } i \text{ can be accomplished at marking } \boldsymbol{m}_j \\ 0, & \text{otherwise;} \quad \forall i = 1, \dots, n, \ \forall j = 0, \dots, n. \end{cases} \tag{6.19}$$

In (6.19) we used "can be accomplished" with the sense that marking $\boldsymbol{m}_j$ means that there are at least $r(i)$ robots in place $g(i)$. We avoid using "is accomplished", because if task i was already served at a previous marking, it cannot be accomplished again at subsequent markings. Additionally, it would be restrictive to avoid more than $r(i)$ robots to be in location $g(i)$ after task i was served (this requirement can be imposed by replacing "$\geq$" with "$=$" in Eq. (6.20)). We need to serve all tasks, and thus we can impose the constraints

$$\sum_{j=0}^{n} x_{i,j} \geq 1, \ \forall i = 1, \dots, n, \tag{6.20}$$

which simply says that we have enough robots in the proper location for serving task i at least at one intermediate marking $\boldsymbol{m}_j$, $j = 0, \dots, n$. As previously stated, if there are more markings satisfying this requirement, the task will be served at the first of such markings, i.e. task i is served at marking $\boldsymbol{m}_j$, with $j = \min\left(\text{argmax}_k \ x_{i,k}\right)$.

Next, we should ensure that the values of binary variables from (6.19) are correctly linked with RMPN markings. For this, first define the characteristic row vector of task i, $i = 1, \dots, n$ (similar to the characteristic vector of output y_i in Section 4.4), as being $\boldsymbol{v}_i \in \{0,1\}^{1 \times |P|}$, with $\boldsymbol{v}_i[k] = 1$ if $k = g(i)$ and $\boldsymbol{v}_i[k] = 0$ for $k \in \{1, \dots, |P|\} \setminus \{g(i)\}$. Thus, the product $\boldsymbol{v}_i \cdot \boldsymbol{m}_j$ gives the number of robots located in place $g(i)$ at marking $\boldsymbol{m}_j$. Since $x_{i,j} \in \{0,1\}$ and the number of tokens of the RMPN model is always equal to $|R|$ (number of robots), the interpretation from Eq. (6.19) is equivalent to the following system of linear inequalities:

$$\begin{cases} \boldsymbol{v}_i \cdot \boldsymbol{m}_j - r(i) \cdot x_{i,j} \geq 0 \\ \boldsymbol{v}_i \cdot \boldsymbol{m}_j - r(i) + 1 \leq (|R| + 1) \cdot x_{i,j} \\ \qquad \forall i = 1, \dots, n; \forall j = 0, \dots, n. \end{cases} \tag{6.21}$$

Informally, the first constraint from (6.21) sets $x_{i,j}$ to 0 if there are less than $r(i)$ robots in place $g(i)$ at marking $\boldsymbol{m}_j$, and leaves it free (either 0 or 1) otherwise. If there are at least $r(i)$ robots in place $g(i)$ at marking $\boldsymbol{m}_j$, the second inequality from (6.21) holds only if $x_{i,j}$ equals 1. For $x_{i,j} = 1$, the right term of (6.21) is a sufficient upper bound for the possible values of the left term, since the number of tokens in the RMPN model is $|R|$ (the RMPN is a live and bounded state machine).

The correct evolution of the RMPN model and the accomplishment of all tasks is enforced by putting together the constraints from (6.18), (6.20), and (6.21), together with correct varying intervals (positive integers for elements of markings and firing vectors, and binary values for witnesses $x_{i,j}$). By imposing as an optimality criterion the total number of fired transitions (as mentioned in Section 6.5.1), the following MILP formulation results:

$$\begin{aligned}
&\min \sum_{j=1}^{n} \mathbf{1}^T \cdot \sigma_j \\
\text{s.t. } & \boldsymbol{m}_j = \boldsymbol{m}_{j-1} + \boldsymbol{C} \cdot \sigma_j \ , \ j = 1, \dots, n \\
& \boldsymbol{v}_i \cdot \boldsymbol{m}_j - r(i) \cdot x_{i,j} \geq 0 \ , \ i = 1, \dots, n, \ j = 0, \dots, n \\
& \boldsymbol{v}_i \cdot \boldsymbol{m}_j - (|R| + 1) \cdot x_{i,j} \leq r(i) - 1 \\
& \sum_{j=0}^{n} x_{i,j} \geq 1 \ , \ i = 1, \dots, n \\
& \boldsymbol{m}_j \in \mathbb{N}_{\geq 0}^{|P|}, \sigma_j \in \mathbb{N}_{\geq 0}^{|T|}, j = 1, \dots, n \\
& x_{i,j} \in \{0, 1\} \ , \ i = 1, \dots, n, \ j = 0, \dots, n.
\end{aligned} \tag{6.22}$$

The optimization problem (6.22) can be solved with existing software tools, e.g. [150, 196]. Although it is an NP-hard problem, we managed to solve fairly complex instances of (6.22) in feasible time. This fact is included in the general remarks that the expected time for obtaining an optimal solution cannot be a priori estimated, and its dependence on various factors (number of constraints, variable bounds, shape of the feasible set) cannot be explicitly formulated.

We use the solution of optimization (6.22) for finding robotic strategies that accomplish all tasks, as in Algorithm 6.8. In RMPN terms, we should obtain the firing sequences σ_j corresponding to the firing count vectors σ_j obtained by solving (6.22). First, let us notice that all markings $\boldsymbol{m}_j$ solutions of (6.22) are reachable markings (and not spurious) since the RMPN is a live state machine by assumption. Second, let us notice that in order to reach markings $\boldsymbol{m}_j$ from $\boldsymbol{m}_{j-1}$ by minimizing $\mathbf{1}^T \cdot \sigma_j$, the optimal solution of each robot does not contain any cycle, i.e. each robot is crossing at most once through each region $p_i \in P$. Therefore, if a robot is in region p_i, only one transition from $p_i^\bullet$ could be fired according to σ_j. Moreover, since the robots are identical, if two robots are in p_i, two output transitions from $p_i^\bullet$ could be fired but it is not important which one is taken by the robots. Based on these remarks, the firing sequences for each robot and for each σ_j are obtained by firing the enabled transitions one by one from $\boldsymbol{m}_{j-1}$ (steps 4 to 9 in Algorithm 6.8).

As stated before, for a given task i we can find at what intermediate marking $\boldsymbol{m}_j$ the task is satisfied (the first time there are enough robots in the required place): $j = \min\left(\text{argmax}_k \ x_{i,k}\right)$ (step 11 in Algorithm 6.8). This gives us the global order

Algorithm 6.8: Iterative construction of agent strategies for task gathering problem

Input: RMPN Q, σ_i, $x_{i,j}$, $i = 1, \dots, n$, $j = 0, \dots, n$
Output: Robot movement strategies

1 Let $\boldsymbol{m} = \boldsymbol{m}_0$;
2 Set all robots to non-waiting state;
3 **for** $j = 1, 2, \dots, n$ **do**
4   **while** $1^T \cdot \sigma_j > 0$ **do**
5   Let $t \in T$ s.t. $\sigma_j[t] > 0 \wedge \boldsymbol{m}[{}^\bullet t] > 0$;
6   Pick any non-waiting robot i in ${}^\bullet t$;
7   Assign movement according to t to robot i;
8   Let $\boldsymbol{m} := \boldsymbol{m} + C[\cdot, t]$;
9   Let $\sigma_j[t] := \sigma_j[t] - 1$;
10   /* At this point $\boldsymbol{m} == \boldsymbol{m}_j$. */
11     $tasks_{to\_serve} = \{i \in \{1, \dots, n\} \mid x_{i,j} == 1 \wedge j == \min\left(\text{argmax}_k\, x_{i,k}\right)\}$
12   **for** $i \in tasks_{to\_serve}$ **do**
13   Set each robot from $g(i)$ to wait until $r(i)$ robots gather in $g(i)$;
14   Serve task i and set all robots from $g(i)$ to the non-waiting state.

in which tasks are served, and it is used for knowing which robots should wait and in what places until enough agents gather there and can accomplish a task (steps 12 to 14 in Algorithm 6.8). Notice that Algorithm 6.8 contains the step from Algorithm 6.1 but additionally constructs the robot strategies for task gathering problems.

After processing the solution of problem (6.22) in Algorithm 6.8, we obtain for each robot a sequence of transitions from T intercalated with waiting modes in place until tasks are served. Based on these sequences, each robot individually moves and we have the guarantee that all tasks will be accomplished. Notice that the waiting modes in place will not introduce any deadlock situation since the order in which those tasks are served is determined at the global level in problem (6.22) and distributed to the robots in Algorithm 6.8. The feasible sequences for the robots are computed by steps 4 and 5 of Algorithm 6.8 while steps 12 to 14 introduce only some delays in the transition firings. Hence, the sequences are predefined and also the order of serving the tasks.

Example 6.4 *Consider the eight regions and three robots in Figure 6.9, modeled by the RMPN in Figure 6.8. All robots are initially deployed in p_8. Assume $n = 4$ tasks to be served given in Example 6.3.*

The MILP formulated as in (6.22) has $n \cdot (|P| + |T|) = 144$ integer variables (for intermediary markings and firing sequences) and $n^2 = 16$ binary variables $x_{i,j}$. The MILP was solved in less than one second and has an optimal cost of 24 transitions to fire for serving all tasks. Algorithm 6.8 yields the following individual robot strategies:

- *Robot 1: t_{24}, serve task 4; t_{12}, t_6, wait until 3 robots in p_2, serve task 2; t_2, wait until 2 robots in p_1, serve task 1;*
- *Robot 2: t_{28}, t_{18}, wait until 2 robots in p_4, serve task 3; t_8, wait until 3 robots in p_2, serve task 2; t_2, wait until 2 robots in p_1, serve task 1;*
- *Robot 3: t_{28}, t_{18}, wait until 2 robots in p_4, serve task 3; t_8, wait until 2 robots in p_2, serve task 2.*

We mention that we successfully solved more demanding examples in feasible time, e.g. on an RMPN with 50 places, 140 transitions, 7 robots, and 4 tasks a solution was obtained in 6 seconds.

6.6 Deadlock prevention using resource allocation models

In previous sections, different strategies to compute robot trajectories based on Petri net models have been presented. However, in the case of multirobot systems a very important problem is to avoid collisions between robots during the motion control. This section explains a different strategy to the one previously proposed in Section 5.7 and is based on a particular Petri net model called RARMPN, defined in Definition 4.5.

Let us assume that the robots are initially located in regions where the collisions can be avoided using local control laws, such as, for example, in [67, 222]. In general, such regions are called depots. Let us assume that the trajectories of robots should start from and finish in the same *depot* region. In order to avoid collisions between robots, the number of robots that can be simultaneously in some regions is limited while the capacity of a depot is equal to the number of robots (unlimited capacity). If the environment is modeled by a graph G and the set of trajectories of robot $r_i \in R$ is denoted as $\mathcal{T}_i$, Algorithm 4.3 computes the corresponding RARMPN.

The technique presented in this section computes *control places* that will be added to the model in order to avoid deadlock situations such as the ones explained in Section 5.7. These control places limit the number of robots in a given set of regions and can be easily implemented using a centralized controller (for example a computer that is able to communicate with all robots in the environment at any time moment). In terms of a Petri net, these control places are called monitor and prevent the *bad siphons* (defined later in this subsection) to be emptied. In order to better understand this, let us consider the following example.

Assume that the environment in Figure 6.10 contains three regions of interest, $O_1 = \{c_{15}\}$, $O_2 = \{c_8\}$ and $O_3 = \{c_{13}\}$ and that there are two robots, both initially located in region c_{16}. Since this is a big area region, we may assume that the collisions between robots cannot occur if both robots are inside it. Identically, we may assume that in regions c_{18} and c_9 robots cannot collide since, again, these regions have big areas and a local controller can be implemented when the robots are inside them.

Let us assume that the team of robots should accomplish the following mission expressed as a boolean formula defined on the outputs y_1, y_2, and y_3 assigned to these regions of interest: $\varphi = !Y1 \ \& \ y2 \ \& \ y3$ with the interpretation given in Section 6.2, i.e. regions O_2 and O_3 should be reached but O_1 should be avoided during the trajectories (for example O_1 is an obstacle and hence always avoided). Moreover, assume that the robots should arrive at O_2 and O_3 and then return to c_{16}.

Using the strategy presented in Section 6.2.4 and implemented in RMTool, the following paths are obtained:

- $r_1 = c_{16}c_{18}c_{19}c_{17}c_9c_2c_6c_8$,
- $r_2 = c_{16}c_{18}c_{19}c_{17}c_9c_{14}c_{13}$.

Since the robots should come back to c_{16}, the trajectory for the first robot is $tr_1 = \{c_{16}\ c_{18}\ c_{19}\ c_{17}\ c_9\ c_2\ c_6\ c_8\ c_6\ c_2\ c_9\ c_{17}\ c_{19}\ c_{18}\ c_{16}\}$ while for the second one is $tr_2 = \{c_{16}\ c_{18}\ c_{19}\ c_{17}\ c_9\ c_{14}\ c_{13}\ c_{14}\ c_9\ c_{17}\ c_{19}\ c_{18}\ c_{16}\}$. Hence, $\mathcal{T}_1 = \{tr_1\}$ and $\mathcal{T}_2 = \{tr_2\}$. By applying Algorithm 4.3 for these sets of trajectories, the RARMPN in Figure 6.11 is obtained.

If P is the set of places of a Petri net, a *trap* is a subset of places ($\Theta \subseteq P$) with the property that the set of output transitions of all places in Θ is contained in (or equal to) the set of input transitions of all places in Θ. For example, $\Theta = \{c_8, R_1c_8\}$ is a trap because its set of output transitions,

$$\Theta^\bullet = {c_8}^\bullet \cup {R_1c_8}^\bullet = \{t_7\} \cup \{t_8\} = \{t_7, t_8\},$$

is included in the set of input transitions,

$$^\bullet\Theta = {^\bullet c_8} \cup {^\bullet R_1c_8} = \{t_8\} \cup \{t_7\} = \{t_7, t_8\}.$$

Formally, $\Theta \subseteq P$ is a trap if $\Theta^\bullet \subseteq {^\bullet\Theta}$. The important property of a trap is that it remains marked if initially it is marked, i.e. if there exists at least one token in a place of the trap at the initial marking then there will always be at least one token in at least one place of the trap. This is due to the fact that if tokens are removed from the trap (by firing an output transition of a place), at least one token is created in the trap (since the set of output transitions is included in the set of input transitions). In our example, because Θ is initially marked (c_8 contains a token), Θ will remain marked for any sequence of transition firings. In this particular case,

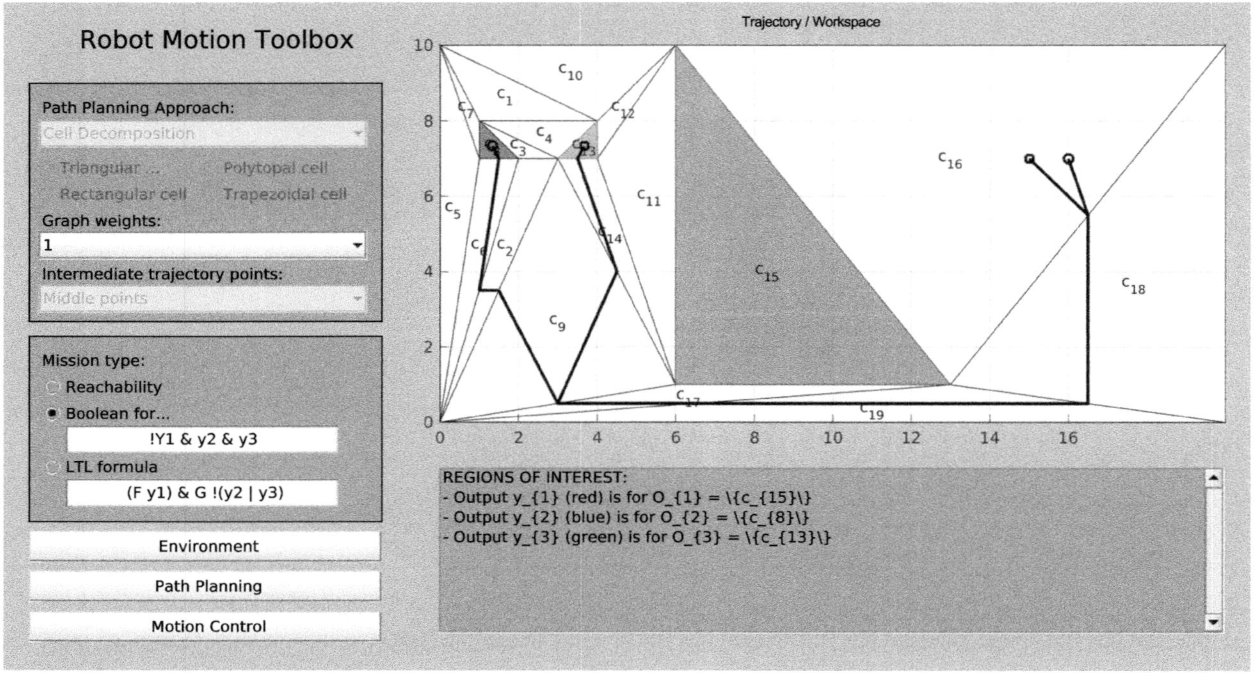

Figure 6.10 Environment with three regions of interest sketched using RMTool.

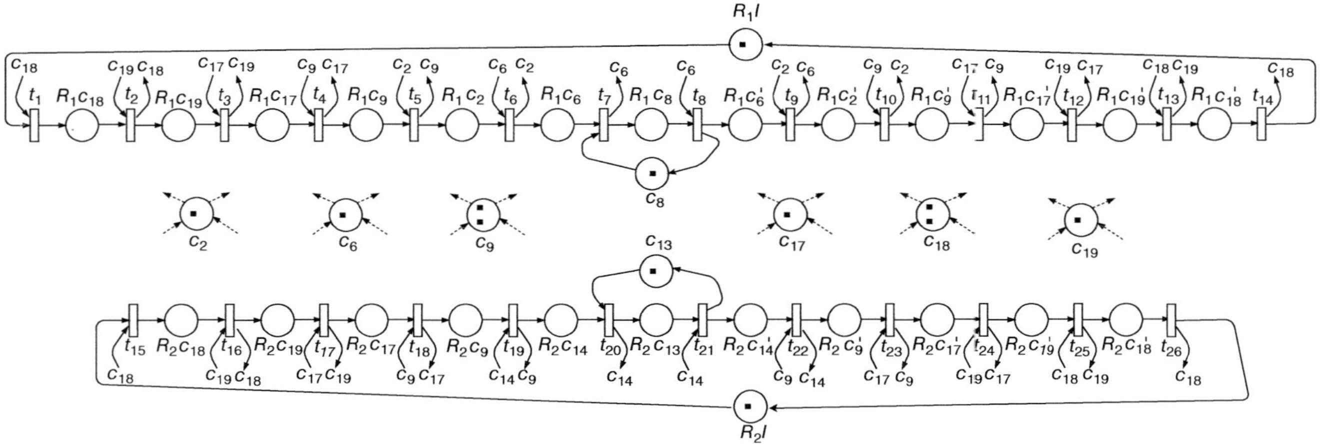

Figure 6.11 Example of a RARMPN model. The RARMPN model of two robots evolving in the environment given in Figure 6.10 that should follow two different paths to reach c_8 and c_{13} respectively and return back to their initial positions. The arcs from/to capacity places c_2, c_6, c_9, c_{17}, c_{18} and c_{19} are shown only omitted for sake of clarity. However, small arcs to denote the start and final nodes are depicted.

the sum of tokens in places of Θ remains constant and equal to one since Θ is also a support of a P-semiflow (but this is not alway true).

On the other hand, a siphon is a subset of places such that its set of input transitions is contained in the set of output transitions. Let us consider the set of places $\Sigma = \{R_1c_{17}, R_1c'_{19}, R_2c_{17}, R_2c'_{19}, c_{17}, c_{19}\}$ in the RARMPN of Figure 6.11. This set of places is a siphon because the set of input transitions,

$$\begin{aligned} {}^{\bullet}\Sigma &= {}^{\bullet}R_1c_{17} \cup {}^{\bullet}R_1c'_{19} \cup {}^{\bullet}R_2c_{17} \cup {}^{\bullet}R_2c'_{19} \cup {}^{\bullet}c_{17} \cup {}^{\bullet}c_{19} = \\ &= \{t_3\} \cup \{t_{12}\} \cup \{t_{19}\} \cup \{t_{26}\} \cup \{t_4, t_{12}, t_{20}, t_{26}\} \cup \{t_3, t_{13}, t_{19}, t_{27}\} = \\ &= \{t_3, t_4, t_{12}, t_{13}, t_{19}, t_{20}, t_{26}, t_{27}\}, \end{aligned}$$

is included in the set of output transitions,

$$\begin{aligned} \Sigma^{\bullet} &= R_1c_{17}{}^{\bullet} \cup R_1c'_{19}{}^{\bullet} \cup R_2c_{17}{}^{\bullet} \cup R_2c'_{19}{}^{\bullet} \cup c_{17}{}^{\bullet} \cup c_{19}{}^{\bullet} = \\ &= \{t_4\} \cup \{t_{13}\} \cup \{t_{20}\} \cup \{t_{27}\} \cup \{t_3, t_{11}, t_{19}, t_{25}\} \cup \{t_2, t_{12}, t_{18}, t_{26}\} = \\ &= \{t_2, t_3, t_4, t_{11}, t_{12}, t_{13}, t_{18}, t_{19}, t_{20}, t_{25}, t_{26}, t_{27}\}. \end{aligned}$$

Formally, $\Sigma \subseteq P$ is a siphon if ${}^{\bullet}\Sigma \subseteq \Sigma^{\bullet}$. A *bad siphon* is a siphon not containing any trap: Σ is a bad siphon if no trap Θ exists such that $\Theta \subseteq \Sigma$. A bad siphon could be emptied during the evolution and the system could reach a deadlock or a livelock state. Since there is a direct relation between deadlock/livelock and the siphons, a possibility to avoid deadlocks is to control all bad siphons and to prevent them from becoming empty. This can be done by adding new control places to the Petri net, one for each bad siphon, prohibiting the bad siphons from losing the last token.

The siphon Σ defined before is a bad siphon and can be emptied by firing the following sequence of transitions: $\sigma = t_1t_2t_3t_4t_5t_6t_7t_8t_9t_{10}t_{11}t_{17}t_{18}$. The marking that is reached after the firing of this sequence corresponds to the following states of the robots: the first robot is in region c_{17} trying to advance to c_{19} but this is not possible because the second robot is in c_{19} trying to advance to c_{17}. Since both c_{17} and c_{19} have capacity one, the system is in a deadlock state.

In order to avoid the emptiness of a bad siphon Σ_k a place p_k called a monitor place can be added such that:

$$\begin{aligned} \mathbf{C}[p_k, t_j] &= \sum_{p_l \in \Sigma_k} \mathbf{C}[p_l, t_j]; \\ \boldsymbol{m}_0[p_k] &= \left(\sum_{p_l \in \Sigma_k} \boldsymbol{m}_0[p_l] \right) - 1. \end{aligned} \tag{6.23}$$

For the bad siphon defined before ($\Sigma = \{R_1c_{17}, R_1c'_{19}, R_2c_{17}, R_2c'_{19}, c_{17}, c_{19}\}$), the monitor place that prevents it from becoming empty is denoted p_k in Figure 6.11. Notice that the output transitions of p_k are transitions $\Sigma^{\bullet} \backslash {}^{\bullet}\Sigma$ (i.e. output transitions of the siphon but not input such that by firing any of these transitions the siphon *is loosing* a token). Marking of place p_k enable or disable the firing of these transitions.

RARMPN has a behavior equivalent to the well-known class of S^4PR used in Resource Allocation Systems [200]. For S^4PR, any deadlock state is associated with at least one minimal empty siphon. Therefore, is important to have algorithms to compute the set of all minimal bad siphons. Two kinds of approaches exist: the first one is based on *Integer Linear Programming* (ILP) [200], while the second one is based on the *pruning graph* [23]. Moreover, in [200] deadlock prevention is solved by an iterative algorithm similar to Algorithm 6.9.

Algorithm 6.9 can be used to compute the set of monitor places that will ensure that the deadlock states will not appear. From an implementation point of view, it can be done by using a centralized controller. When a robot should enter a region by firing a transition that is the output transition of a monitor place, it should check if the monitor place has enough tokens. Moreover, the capacity constraints can be implemented by using local sensors since it is necessary only to check if in the next region that it should enter there exists *enough space*. However, this can be implemented in the same centralized controller where the monitor places are implemented.

By applying Algorithm 6.9 to the RARMPN in Figure 6.11, the only monitor place that should be considered and implemented is place p_k in the Figure 6.12. Notice that not all monitor places should be implemented. It could happen that a bad siphon controlled by a monitor place is composed of many capacity places and the initial marking of the monitor place results, according to (6.23), is found to be greater than the number of robots. Obviously, if this is the case, the monitor place

---

Algorithm 6.9: Liveness enforcement

Input: RARMPN model that may contain empty siphons.
Output: Live RARMPN.

Compute the set S of minimal bad siphons of the input RARMPN model
while $S \neq \emptyset$ **do**
    forall $\Sigma_i \in S$ **do**
        $P_C = P_C \cup \{p_i\}$; `/* add a monitor` p_i `to control` S_i `*/`
        forall $t_j \in T$ **do**
            $\boldsymbol{C}[p_i, t_j] = \sum_{p_l \in \Sigma_i} \boldsymbol{C}[p_l, t_j]$;
        $\boldsymbol{m}_0[p_i] = \left(\sum_{p_l \in \Sigma_i} \boldsymbol{m}_0[p_l] \right) - 1$;
        Compute the set S of minimal bad siphons for the new RARMPN model

---

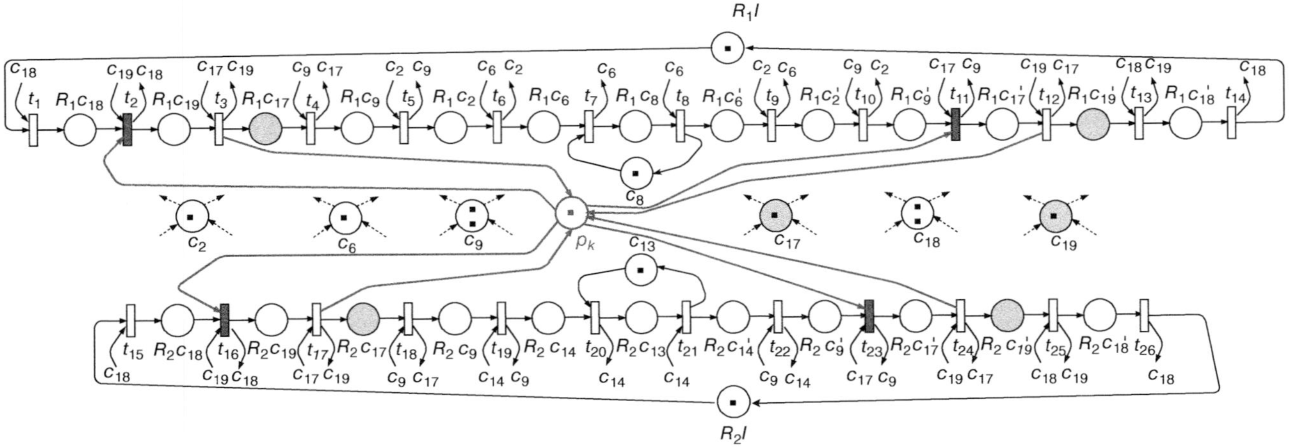

Figure 6.12 Example of a RARMPN model with a monitor place p_k. The RARMPN model with the monitor place p_k that is controlling the bad siphon $\Sigma = \{R_1c_{17}, R_1c'_{19}, R_2c_{17}, R_2c'_{19}, c_{17}, c_{19}\}$ preventing it to be emptied. Yellow places compose the bad siphons, red transitions are output (but not input) transitions of the siphon while red place p_k is the monitor place.

is an *implicit place*[4] and it is not necessary to be implemented. This could happen because the bad siphons are structural elements (they are computed based on the structure without considering the initial marking of the net system).

6.7 Conclusions

Robot Motion Petri net (RMPN) models have been used in this chapter for solving various path planning problems. The requirements are given as global tasks for the whole team of identical robots, i.e. without specific robot-to-task assignments. The RMPN model maintains a fixed topology with respect to the number of robots, rather than strategies based on synchronous products of transition system models (as in Chapter 5). The solutions are included in the mathematical programming branch from operations research, since they use mixed integer linear programming optimization problems for finding a movement plan for each robot. The first presented problem is for the case of team requirements expressed as Boolean-based formulas (Section 6.2), where the robots can follow the designed paths without synchronizing among themselves. Section 6.3 assumes LTL requirements, and the solution is iteratively constructed by solving multiple optimization problems, after which the robots should follow their trajectories and synchronize in specific points. Section 6.4 assumes a team requirement where each robot has to collect a number of resources in an imposed order, and the formulated optimization problem belongs in this case to the less complex linear programming class. Different from previous requirements, Section 6.5 considers some task locations in the environment that can be accomplished (served) only by having more robots in the same position. After robotic paths are obtained, Section 6.6 presents deadlock prevention methods based on constructing Resource Allocation Robot Motion PN (RARMPN) models.

4 An *implicit place* is a place that is never the only one that is constraining the firing of transitions. In other words, if removed, the behavior of the net system does not change [192].

7

Concluding Remarks

In this monograph, we have studied the problem of path planning of cooperative mobile robots by using the paradigm of discrete-event systems. Every chapter contributes a particular aspect within this context. In particular, we have exploited the advantages of Boolean Logic, Linear Temporal Logic, cell decomposition, Finite State Automata modeling, and Petri Nets to propose novel approaches that can be used to generate routes for a team of robots operating in cluttered environments and achieving a common high-level specification.

In addition to the substantial theoretical contribution, this book also contributed an interactive software tool RMTool, running in MATLAB, in which a user can simulate the strategies introduced and formulated here. This adds another major value to this book. The user can interact with the tool and understand much more easily the proposed theoretical contributions, as well as use some provided functions to build new algorithms.

The relevance of these contributions is also supported by the publication of a substantial part of the contents of this monograph in renowned journals and conferences in the fields of robotics and discrete-event systems.

The authors would like to express future directions to further expand the contents of this monograph. This future work could be divided into theoretical contributions, practical contributions, and contributions related to the software tool proposed here. Regarding the first area, path planning for mobile robots in (partially) unknown environments is one of the main interesting extensions that will be considered in the future. Furthermore, extensions for team of non-identical robots is also one of the interesting directions for theoretical contributions. In relation to the practical contributions, the authors have already started the process of testing the algorithms and programs included in this book in real mobile robots operating in real conditions. This step will reinforce the suitability and robustness of the theoretical concepts explained here. Finally, the authors plan to improve

Path Planning of Cooperative Mobile Robots Using Discrete Event Models, First Edition.
Cristian Mahulea, Marius Kloetzer, and Ramón González.

the capabilities of the software tool proposed here. These improvements will cover the integration with more powerful robotics-related environments such as Robot Operating System (ROS) with Gazebo or Webots (Cyberbotics).

Bibliography

1 CGAL, *Computational Geometry Algorithms Library*. http://www.cgal.org.

2 S. Agarwal, A.K. Gaurav, M.K. Nirala, and S. Sinha. Potential and sampling based rrt star for real-time dynamic motion planning accounting for momentum in cost function. In L. Cheng, A. Leung, and S. Ozawa, editors, *Neural Information Processing*, volume 11307 of *Lecture Notes in Computer Science*. Springer, 2018.

3 R. Alur, C. Courcoubetis, and D.L. Dill. Model-checking for real-time systems. In *IEEE Symposium on Logic in Computer Science*, pages 414–425, Philadelphia, PA, 1990.

4 O. Amidi. *Integrated Mobile Robot Control*. Technical Report CMU-RI-TR-90-17, Robotics Institute - Carnegie Mellon University, Pittsburgh, PA, USA, 1990.

5 D. Apostolopoulos, L. Pedersen, B. Shamah, K. Shillcutt, M.D. Wagner, and W.R. Whittaker. *Robotic antarctic meteorite search: Outcomes*. pages 4174–4179. *IEEE International Conference on Robotics and Automation*, IEEE, 2001.

6 R. Arkin and R. Murphy. Autonomous navigation in a manufacturing environment. *IEEE Transactions on Robotics and Automation*, 6(4): 445–454, 1990.

7 R.C. Arkin. *Designing Autonomous Agents*, chapter Integrating behavioral, perceptual and world knowledge in reactive navigation, pages 105–122. MIT Press, Cambridge, MA, 1990.

8 K.O. Arras. *The CAS Robot Navigation Toolbox: Users Guide and Reference*. Center for Autonomous Systems, KTH, September 2004.

9 K.J. Aström and R.M. Murray. *Feedback Systems: An Introduction for Scientists and Engineers*. Princeton University Press, USA, 2008.

10 C. Baier and J.P. Katoen. *Principles of Model Checking*. The MIT Press, USA, 2008.

Path Planning of Cooperative Mobile Robots Using Discrete Event Models, First Edition.
Cristian Mahulea, Marius Kloetzer, and Ramón González.

11 F.A. Bayer, F.D. Brunner, M. Lazar, M. Wijnand, and F. Allgöwer. *A tube-based approach to nonlinear explicit MPC*. In *2016 IEEE 55th Conference on Decision and Control (CDC 2016), December 12-14, 2016. Las Vegas, USA*, pages 4059–4064, 2016.

12 C. Belta and L.C.G.J.M. Habets. Controlling a class of nonlinear systems on rectangles. *IEEE Transactions on Automatic Control*, 51(11): 1749–1759, 2006.

13 C. Belta, A. Bicchi, M. Egerstedt, E. Frazzoli, E. Klavins, and G.J. Pappas. Symbolic Planning and Control of Robot Motion. *IEEE Robotics and Automation Magazine*, 14(1):61–71, 2007.

14 C. Belta, B. Yordanov, and E.A. Gol. *Formal Methods for Discrete-Time Dynamical Systems*. Springer International Publishing, 2017.

15 M. De Berg, O. Cheong, and M. van Kreveld. *Computational Geometry: Algorithms and Applications*. Springer, 3rd edition, 2008.

16 M.U. Bers. Coding, playgrounds and literacy in early childhood education: The development of kibo robotics and scratchjr. In *IEEE Global Engineering Education Conference*, pages 2094–2102, Tenerife, 2018.

17 A.M. Bloch, J.E. Marsden, and D.V. Zenkov. Nonholonomic dynamics. *Notices of the American Mathematical Society*, 52(3):320–329, 2005.

18 S. Boyd and L. Vandenberghe. *Convex Optimization*. Cambridge University Press, New York, USA, 2004.

19 F.M. Brown. *Boolean Reasoning: The Logic of Boolean Equations*. Dover Publications, 2nd edition, 2012.

20 A. F. Browne and J. M. Conrad. A versatile approach for teaching autonomous robot control to multi-disciplinary undergraduate and graduate students. *IEEE Access*, 6:25060–25065, 2018.

21 C. Buiu. Hybrid Educational Strategy for a Laboratory Couse on Cognitive Robotics. *IEEE Transactions on Education*, 51(1):100–107, 2008.

22 G. Campion, G. Bastin, and B. D'Andréa-Novel. Structural Properties and Classification of Kinematic and Dynamic Models of Wheeled Mobile Robots. *IEEE Transactions on Robotics and Automation*, 12(1):47–62, 1996.

23 E.E. Cano, C.A. Rovetto, and J.M. Colom. An algorithm to compute the minimal siphons in S^4PR nets. *Discrete Event Dynamic Systems*, 22(4):403–428, December 2012.

24 C. Canudas, B. Siciliano, and G. Bastin. *Theory of Robot Control*. Communications and Control Engineering. Springer, Germany, 2nd edition, 1997.

25 T. Cao and A.C. Sanderson. Task decomposition and analysis of robotic assembly task plans using petri nets. *IEEE Trans. on Industrial Electronics*, 41(6):620–630, 1994.

26 S. Carpin, M. Lewis, W. Jijun, and S. Balakirsky. USARSim: A Robot Simulator for Research and Education. In *IEEE Int. Conf. on Robotics and Automation (ICRA)*, pages 1400–1405. IEEE, April 10-14, 2007.

27 C. Cassandras and S. Lafortune. *Introduction to Discrete Event Systems.* Springer-Verlag New York, Inc., Secaucus, NJ, USA, 2nd edition, 2006.

28 L. Champeny-Bares, S. Coppersmith, and K. Dowling. *The terregator mobile robot.* Techical Report CMU-RI-TR-93-03, The Robotics Institute, Carnegie Mellon University, Pittsburgh, PA, USA, 1991.

29 C. Chang, L. Wenyin, and H. Zhang. Image retrieval based on region shape similarity. In *Society of Photo-Optical Instrumentation Engineers (SPIE) Conference Series*, volume 4315, pages 31–38, 2001.

30 D. Chaos, J. Chacon, J.A. Lopez-Orozco, and S. Dormido. Virtual and Remote Robotic Laboratory Using EJS, Matlab and LabVIEW. *Sensors*, 13(2):2595–2612, 2013.

31 B. Chazelle. Approximation and decomposition of shapes. In J.T. Schwartz and C.K. Yap, editors, *Algorithmic and Geometric Aspects of Robotics*, pages 145–185. Lawrence Erlbaum Associates, Hillsdale, NJ, 1987.

32 Y. Chen, Z. Li, and M. Zhou. Optimal Supervisory Control of Flexible Manufacturing Systems by Petri Nets: A Set Classification Approach. *IEEE Trans. Automation Science and Engineering*, 11(2):549–563, 2014.

33 L. P. Chew. *Constrained delaunay triangulations.* In *Proceedings of the Third Annual Symposium on Computational Geometry*, SCG '87, pages 215–222, 1987.

34 J.W. Chinneck. *Practical Optimization: A Gentle Introduction.* Available at http://www.sce.carleton.ca/faculty/chinneck/po.html, 2004.

35 H. Choset, K. M. Lynch, S. Hutchinson, G. Kantor, W. Burgard, L.E. Kavraki, and S. Thrun. *Principles of Robot Motion: Theory, Algorithms, and Implementations.* MIT Press, Boston, 2005.

36 A. Cimatti, E.M. Clarke, F. Giunchiglia, and M. Roveri. *NUSMV: A New Symbolic Model Verifier.* In *Proceedings of the 11th International Conference on Computer Aided Verification*, CAV '99, pages 495–499, London, UK, 1999. Springer-Verlag.

37 E. Clarke, D. Peled, and O. Grumberg. *Model Checking.* MIT Press, 1999.

38 X. Cong, C. Gu, M. Uzam, Y. Chen, A.M. Al-Ahmari, N. Wu, M. Zhou, and Z. Li. Design of Optimal Petri Net Supervisors for Flexible Manufacturing Systems via Weighted Inhibitor Arcs. *Asian Journal of Control*, 20(1):511–530, 2018.

39 J.D. Contreras, F. Martínez, and F.H. Martinez. Path planning for mobile robots based on visibility graphs and A* algorithm. In *Seventh International Conference on Digital Image Processing (ICDIP)*. SPIE, July 2015.

40 P. Corke. A Robotics Toolbox for Matlab. *IEEE Robotics & Automation Magazine*, 3(1):24–32, 1996.

41 P. Corke. *Robotics, Vision and Control. Fundamental algorithms in Matlab.* Springer Tracts in Advanced Robotics. Springer, Germany, 2011.

42 T.H. Cormen, C.E. Leiserson, R.L. Rivest, and C. Stein. *Introduction to Algorithms.* The MIT Press and McGraw-Hill Book Company, Cambridge, Massachusetts and New York, 2nd edition, 2001.

43 H. Costelha and P. Lima. Robot task plan representation by Petri nets: modelling, identification, analysis and execution. *Journal of Autonomous Robots*, pages 1–24, 2012.

44 M.S. Couceiro. *MRSim - Multi-Robot Simulator.* Available at http://www2.isr.uc.pt/micaelcouceiro/media/help/helpMRSim.htm, 2012.

45 R.C. Coulter. *Implementation of the pure pursuit path tracking algorithm.* Technical Report CMU–RI–TR–92–01, Robotics Institute, Carnegie Mellon University, USA, January 1992.

46 A. Cruz-Martin, J.A. Fernandez-Madrigal, C. Galindo, J. Gonzalez-Jimenez, C. Stockmans-Daou, and J.L. Blanco-Claraco. A Lego Mindstorms NXT Approach for Teaching at Data Acquisition, Control Systems Engineering and Real-Time Systems Undergraduate Courses. *Computers and Education*, pages 1–49, 2012.

47 E. Dallal, A. Colombo, D. Del-Vecchio, and S. Lafortune. Supervisory control for collision avoidance in vehicular networks using discrete event abstractions. *Discrete Event Dynamic Systems*, pages 1–44, 2016.

48 G. Dantzig and M. Thapa. *Linear Programming 2: Theory and Extensions.* Spinger, 2003.

49 J.M. Gómez de Gabriel, A. Mandow, J. Fernandez-Lozano, and A.J. Garcia-Cerezo. Using Lego NXT Mobile Robots with Labview for Undergraduate Courses on Mechatronics. *IEEE Transactions on Education*, 54(1):41–47, 2011.

50 J. DeCastro, R. Ehlers, M. Rungger, A. Balkan, and H. Kress-Gazit. Automated generation of dynamics-based runtime certificates for high-level control. *Discrete Event Dynamic Systems*, 27(2):371–405, June 2017.

51 E. Dickmanns. Vehicles capable of dynamic vision. In *Proc. of the Fifteenth International Joint Conference on Artifical Intelligence*, volume 2, pages 1577–1592, 1997.

52 E.D. Dickmanns and A. Zapp. Autonomous high speed road vehicle guidance by computer vision. In *IFAC World Congress*, pages 221–226, Munich, Germany, 1987.

53 E.W. Dijkstra. A note on two problems in connexion with graphs. In *Numerische Mathematik*, volume 1, pages 269–271. Mathematisch Centrum, Amsterdam, The Netherlands, 1959.

54 X. Ding, M. Kloetzer, Y. Chen, and C. Belta. Automatic deployment of robotic teams. *IEEE Robotics and Automation Magazine*, 18(3):75–86, 2011.

55 Z. Ding, H. Qiu, R. Yang, C. Jiang, and M. Zhou. *Interactive-control-model for human-computer interactive system based on petri nets. IEEE Transactions on Automation Science and Engineering*, pages 1–14, 2019. DOI: 10.1109/TASE.2019.2895507.

56 S. Dormido. Control Learning: Present and Future. *Annual Reviews in Control*, 28(1):115–136, 2004.

57 F. Duchon, A. Babinec, M. Kajan, P. Beno, M. Florek, T. Fico, and L. Jurisica. Path Planning with Modified A Star Algorithm for a Mobile Robot. *Procedia Engineering*, 96:59–69, 2014.

58 G. Dudek and M. Jenkin. *Computational Principles of Mobile Robotics*. Cambridge University Press, United Kingdom, 2nd edition, 2010.

59 A. Duret-Lutz, A. Lewkowicz, A. Fauchille, T. Michaud, E. Renault, and L. Xu. *Spot 2.0 - a framework for LTL and ω-automata manipulation*. In *Proc. of ATVA'16*, number 9938 in LNCS, pages 122–129, 2016.

60 M.G. Earl and R. D'Andrea. Iterative MILP methods for vehicle-control problems. *IEEE Trans. Robotics*, 21(6):1158–1167, 2005.

61 E.A. Emerson. Temporal and modal logic. In J. van Leeuwen, editor, *Handbook of Theoretical Computer Science: Formal Models and Semantics*, volume B, pages 995–1072. MIT Press, 1990.

62 J. Esparza and K. Heljanko. *Implementing LTL model checking with net unfoldings*. In *8th international SPIN workshop on Model checking of software*, pages 37–56, 2001. Springer-Verlag New York, Inc.

63 J. Esparza and C. Schrter. Net reductions for LTL model-checking. In T. Margaria and T. Melham, editors, *Correct Hardware Design and Verification Methods*, volume 2144 of *Lecture Notes in Computer Science*, pages 310–324. Springer, Berlin Heidelberg, 2001.

64 J. Ezpeleta, F. García-Vallés, and J-M. Colom. *A class of well structured Petri nets for flexible manufacturing systems*. In Jörg Desel and Manuel Silva, editors, *Application and Theory of Petri Nets 1998, 19th International Conference, ICATPN '98*, volume 1420 of *Lecture Notes in Computer Science*, pages 64–83. Springer, 1998.

65 G.E. Fainekos, H. Kress-Gazit, and G.J. Pappas. Hybrid controllers for path planning: a temporal logic approach. In *IEEE Conference on Decision and Control*, pages 4885–4890, Seville, Spain, December 2005.

66 G.E. Fainekos, A. Girard, H. Kress-Gazit, and G.J. Pappas. Temporal logic motion planning for dynamic robots. *Automatica*, 45(2):343–352, 2009.

67 I. Filippidis, D.V. Dimarogonas, and K.J. Kyriakopoulos. *Decentralized multi-agent control from local ltl specifications*. In *51st IEEE Conference on Decision and Control (CDC)*, pages 6235–6240, 2012.

68 R.A. Finkel and J.L. Bentley. Quad-trees: a data structure for retrieval on composite keys. *ACTA Informatica*, 4:1–9, 1974.

69 C. Floudas and P. Pardolos. *Encyclopedia of Optimization, volume 2*, 2nd edition. Spinger, New York, 2009.

70 S. Fortune. Voronoi diagrams and delaunay triangulations. In D.Z. Du and F. Hwang, editors, *Computing in Euclidean Geometry*, Lecture Notes Series on Computing, pages 193–233. World Scientific, 1992.

71 E. Freund and H. Hoyer. Real-time pathfinding in multirobot systems including obstacle avoidance. *The International Journal of Robotics Research*, 7(1):42–70, 1988.

72 K. Fukuda. *CDD/CDD+ package*. Available at https://www.inf.ethz.ch/personal/fukudak/cdd_home/.

73 P. Gastin and D. Oddoux. Fast LTL to Büchi automata translation. In *Lecture Notes in Computer Science*, volume 2102, pages 53–65. Springer-Verlag, London, UK, 2001.

74 P. Gastin and D. Oddoux. *Fast ltl to büchi automata translation*. In *Proc. of the 13th Conference on Computer Aided Verification (CAV'01)*, number 2102 in LNCS, pages 53–65, 2001.

75 R. Geraerts and M.H. Overmars. A comparative study of probabilistic roadmap planners. In J.D. Boissonnat, J. Burdick, K. Goldberg, and S. Hutchinson, editors, *Algorithmic Foundations of Robotics V*, volume 7, pages 43–58. Springer, Berlin, Germany, 2004.

76 N. Ghita and M. Kloetzer. Trajectory Planning for a Car-like Robot by Environment Abstraction. *Robotics and Autonomous Systems*, 60(4): 609–619, 2012.

77 J. Gómez and E.F. Camacho. Mobile Robot Navigation in a Partially Structured Static Environment, Using Neural Predictive Control. *Control Engineering Practice*, 4(12):1669–1679, 1996.

78 R. Gonzalez. *Contributions to Modelling and Control of Mobile Robots in Off-Road Conditions*. PhD Thesis, University of Almeria, Almeria, Spain, 2011.

79 R. Gonzalez, F. Rodriguez, J. Sanchez-Hermosilla, and J.G. Donaire. Navigation Techniques for Mobile Robots in Greenhouses. *Applied Engineering in Agriculture*, 25(2):153–165, 2009.

80 R. González, M. Fiacchini, J. L. Guzmán, T. Álamo, and F. Rodríguez. Robust Tube-based Predictive Control for Mobile Robots in Off-Road Conditions. *Robotics and Autonomous Systems*, 59(10):711–726, 2011.

81 R. Gonzalez, F. Rodriguez, and J.L. Guzman. *Autonomous Tracked Robots in Planar Off-Road Conditions. Modelling, Localization and Motion Control.* Series: Studies in Systems, Decision and Control. Springer, Germany, 2014.

82 R. Gonzalez, C. Mahulea, and M. Kloetzer. A Matlab-based interactive simulator for mobile robotics. In *IEEE Int. Conf. on Automation Science and Engineering*, Gothenburg, Sweden, 2015.

83 R. Gonzalez, S. Byttner, and K. Iagnemma. *Comparison of Machine Learning Approaches for Soil Embedding Detection of Planetary Exploratory Rovers.* In *8th ISTVS Amercias Conference*, Detroit, MI, USA, September 12-14, 2016. International Society for Vehicle-Terrain Systems.

84 R. Gonzalez, P. Jayakumar, and K. Iagnemma. Generation of stochastic mobility maps for large-scale route planning of ground vehicles: a case study. *Journal of Terramechanics*, 69:1–11, 2017.

85 R. Gonzalez, M. Kloetzer, and C. Mahulea. *Comparative study of trajectories resulted from cell decomposition path planning approaches.* In *2017 21st International Conference on System Theory, Control and Computing (ICSTCC)*, pages 49–54, 2017.

86 I. Griva, S. Nash, and A. Sofer. *Linear and Nonlinear Optimization.* Society for Industrial Mathematics, 2008. ISBN 978-0-89871-661-0. 2nd ed.

87 J.P. Grotzinger, J. Crisp, A.R. Vasavada, C.J. Baker R.C. Anderson, R. Barry, and D.F. Blake. Mars Science Laboratory Mission and Science Investigation. *Space Science Reviews*, 170(1):5–56, 2012.

88 M. Guo, J. Tumova, and D. Dimarogonas. Cooperative decentralized multi-agent control under local LTL tasks and connectivity constraints. In *IEEE Conference on Decision and Control*, pages 75–80, IEEE, Los Angeles, CA, USA, December 2014.

89 J.L. Guzmán, M. Berenguel, F. Rodríguez, and S. Dormido. An Interactive Tool for Mobile Robot Motion Planning. *Robotics and Autonomous Systems*, 56(5):396–409, 2008.

90 L.C.G.J.M. Habets and J.H. van Schuppen. A control problem for affine dynamical systems on a full-dimensional polytope. *Automatica*, 40:21–35, 2004.

91 L.C.G.J.M. Habets, P.J. Collins, and J.H. van Schuppen. Reachability and control synthesis for piecewise-affine hybrid systems on simplices. *IEEE Transactions on Automatic Control*, 51:938–948, 2006.

92 O. Hachour. The proposed hybrid intelligent system for path planning of intelligent autonomous systems. *International Journal of Mathematics and Computers in Simulation*, 3:133–145, 2009.

93 I. Heller and C.-B. Tompkins. *An extension of a theorem of Dantzig's.* In Kuhn and Tucker [128], pages 247–254.

94 D. Helmick, S. Roumeliotis, Y. Cheng, D. Clouse, M. Bajracharya, and L. Matthies. Slip-compensated Path Following for Planetary Exploration Rovers. *Advanced Robotics*, 20(11):1257–1280, 2006.

95 M. Hernandez-Ordonez, M.A. Nuno-Maganda, O. Montano-Rivas C.A. Calles-Arriaga, and K.E. Bautista. An education application for teaching robot arm manipulator concepts using augmented reality. *Mobile Information Systems*, pages 1–8, 2018.

96 A. Hoffman and J. Kruskal. *Integral boundary points of convex polyhedra.* In Kuhn and Tucker [128], pages 223–246.

97 G.J. Holzmann. *The SPIN Model Checker: Primer and Reference Manual*, volume 1003. Addison-Wesley, 2004.

98 J. Hrabec. *Autonomous Mobile Robotics Toolbox SIMROBOT*. Master's Thesis, Department of Control, Measurement and Instrumentation. Brno University of Technology, Brno, Czech Republic, 2001.

99 H. Hu, M. Zhou, and Z. Li. Liveness and Ratio-Enforcing Supervision of Automated Manufacturing Systems Using Petri Nets. *IEEE Trans. Systems, Man, and Cybernetics, Part A*, 42(2):392–403, 2011.

100 H.P. Huang and S.Y. Chung. *Dynamic visibility graph for path planning.* In *IEEE Int. Conf. on Intelligent Robots and Systems (IROS)*, Sendai, Japan, September-October 2004. IEEE.

101 K. Iagnemma and S. Dubowsky. *Mobile Robots in Rough Terrain. Estimation, Motion Planning, and Control with Application to Planetary Rovers.* Springer Tracts in Advanced Robotics. Springer, Germany, 2004. ISBN 9783540219682.

102 IBM. *IBM ILOG CPLEX Optimization Studio.* Software, 2016.

103 K. Jensen and L. Kristensen. *Coloured Petri Nets: Modelling and Validation of Concurrent Systems.* Springer Publishing Company, Incorporated, 1st edition, 2009.

104 J.H. Jeon, S. Karaman, and E. Frazzoli. Anytime Computation of Time-Optimal Off-Road Vehicle Maneuvers using the RRT*. In *IEEE Conf. on Decision and Control*, pages 3276–3282, Orlando, FL, USA, December 2011. IEEE.

105 Y. Kanayama, Y. Kimura, F. Miyazaki, and T. Noguchi. *A Stable Tracking Control Method for an Autonomous Mobile Robots.* pages 384–389. *IEEE International Conference on Robotics and Automation*, IEEE, 1990. Cincinnati, USA.

106 L.E. Kavraki, P. Svestka, J.C. Latombe, and M.H. Overmars. Probabilistic roadmaps for path planning in high-dimensional configuration spaces. *IEEE Transactions on Robotics and Automation*, 12(4):566–580, 1996.

107 K. Klai, S. Haddad, and J.M. Ilie. Modular Verification of Petri Nets Properties: A Structure-Based Approach. In F. Wang, editor, *Formal Techniques for Networked and Distributed Systems - FORTE 2005*, volume 3731 of *Lecture Notes in Computer Science*, pages 189–203. Springer, Berlin Heidelberg, 2005.

108 G. Klancar and I. Skrjanc. Tracking-error Model-based Predictive Control for Mobile Robots in Real Time. *Robotics and Autonomous Systems*, 55(1):460–469, 2007.

109 M. Kloetzer. *Symbolic Motion Planning and Control*. PhD Thesis, College of Engineering, Boston University, USA, 2008.

110 M. Kloetzer and C. Belta. A framework for automatic deployment of robots in 2D and 3D environments. In *IEEE/RSJ International Conference on Intelligent Robots and Systems*, pages 953–958, Beijing, China, 2006.

111 M. Kloetzer and C. Belta. A fully automated framework for control of linear systems from temporal logic specifications. *IEEE Transactions on Automatic Control*, 53(1):287–297, 2008.

112 M. Kloetzer and C. Belta. Automatic deployment of distributed teams of robots from temporal logic motion specifications. *IEEE Transactions on Robotics*, 26(1):48–61, 2010. ISSN 1552-3098.

113 M. Kloetzer and N. Ghita. *Software tool for constructing cell decompositions*. In *Proceedings of the IEEE Conference on Automation Science and Engineering*, pages 507–512, 2011.

114 M. Kloetzer and N. Ghita. *Software tool for constructing cell decompositions*. In *IEEE Conference on Automation Science and Engineering*, pages 507–512, 2011.

115 M. Kloetzer and C. Mahulea. A Petri Net based Approach for Multi-Robot Path Planning. *Discrete Event Dynamic Systems*, 24(2): 417–445, 2014.

116 M. Kloetzer and C. Mahulea. *An assembly problem with mobile robots*. In *IEEE Emerging Technology and Factory Automation (ETFA)*, pages 1–7, 2014.

117 M. Kloetzer and C. Mahulea. *Accomplish multi-robot tasks via petri net models*. In *IEEE International Conference on Automation Science and Engineering (CASE)*, pages 304–309, 2015.

118 M. Kloetzer and C. Mahulea. LTL-Based Planning in Environments with Probabilistic Observations. *IEEE Transactions on Automation Science and Engineering*, 12(4):1407–1420, 2015.

119 M. Kloetzer and C. Mahulea. *Multi-robot path planning for syntactically co-safe LTL specifications*. In *13th International Workshop on Discrete Event Systems (WODES)*, pages 452–458, 2016.

120 M. Kloetzer and C. Mahulea. *Path planning for robotic teams based on LTL specifications and Petri net models*, 2019, under review.

121 M. Kloetzer, C. Mahulea, C. Belta, and M. Silva. An Automated Framework for Formal Verification of Timed Continuous Petri Nets. *IEEE Transactions on Industrial Informatics*, 6(3):460–471, 2010.

122 M. Kloetzer, X.-C. Ding, and C. Belta. *Multi-robot deployment from LTL specifications with reduced communication*. In *IEEE Conference on Decision* *and Control and European Control Conference (CDC-ECC)*, pages 4867–4872, 2011.

123 M. Kloetzer, C. Mahulea, and J.-M. Colom. *Petri net approach for deadlock prevention in robot planning*. In *IEEE 18th Conference on Emerging Technologies Factory Automation (ETFA)*, 2013.

124 M. Kloetzer, C. Mahulea, and R. Gonzalez. Optimizing cell decomposition path planning for mobile robots using different metrics. In *ICSTCC'2015: 19th International Conference on System Theory, Control and Computing*, pages 565–570, Cheile Gradistei, Romania, 2015.

125 S. Koenig and M. Likhachev. Fast replanning for navigation in unknown terrain. *Transactions on Robotics*, 21(3):354–363, 2005.

126 M.K. Kozlov, S.P. Tarasov, and L.G. Khachiyan. The polynomial solvability of convex quadratic programming. *{USSR} Computational Mathematics and Mathematical Physics*, 20(5):223–228, 1980.

127 H. Kuhn and A. Tucker, editors. *Linear Inequalities and Related Systems*. Princeton, NJ: Princeton Univ. Press, 1956.

128 A.N. Kumar. Three Years of using Robots in an Artificial Intelligence Course - Lessons Learned. *ACM Journal on Educational Resources in Computing*, 4(3):1–15, 2004.

129 R. Kumar and V. K. Garg. *Modeling and Control of Logical Discrete Event Systems*. Kluwer, Boston, MA, 1995.

130 B. Lacerda and P.-U. Lima. *LTL-based decentralized supervisory control of multi-robot tasks modelled as Petri nets*. In *IEEE/RSJ Int. Conf. on Intelligent Robots and Systems*, pages 3081–3086, 2011.

131 L. Lamport. The temporal logic of actions. *ACM Transactions on Programming Languages and Systems*, 16(3):872–923, 1994.

132 J. C. Latombe. *Robot Motion Planning*. Kluger Academic Pub., 1991.

133 J.P. Laumond, editor. *Robot Motion Planning and Control*, volume 229 of *Lecture Notes in Control and Information Sciences*. Springer, Berlin, Germany, 1998.

134 S. M. LaValle. *Rapidly-Exploring Random Trees: A New Tool for Path Planning*. Technical Report TR 98-11, Computer Science Dept., Iowa State University, October 1998.

135 S.M. LaValle. *Planning Algorithms*. Cambridge, 2006. Available at http://planning.cs.uiuc.edu.

136 S.U. Lee, R. Gonzalez, and K. Iagnemma. *Robust sampling-based motion planning for autonomous tracked vehicles in deformable high slip terrain*. In *IEEE Int. Conf. on Robotics and Automation (ICRA)*, pages 2569–2574, Stockholm, Sweden, May 16-21, 2016.

137 R. Lenain, B. Thuilot, C. Cariu, and P. Martinet. *Model Predictive Control for Vehicle Guidance in Presence of Sliding: Application to Farm Vehicles Path*

Tracking. pages 885–890. IEEE International Conference on Robotics and Automation, IEEE, April 2005. Barcelona, Spain.

138 M. Li, Q. Sun, Q. Song, Z. Wang, and Y. Li. Path Planning of Mobile Robot Based on RRT in Rugged Terrain. In *International Conference on Computer Science and Application Engineering*, ACM, New York, NY, USA, 2018.

139 Z. Li, N. Wu, and M. Zhou. Deadlock Control of Automated Manufacturing Systems Based on Petri Nets - A Literature Review. *IEEE Trans. Systems, Man, and Cybernetics, Part C*, 42(4):437–462, 2012.

140 M. Lin, K. Yuan, C. Shi, and Y. Wang. Path planning of mobile robot based on improved A* algorithm. In *Chinese Control And Decision Conference*, pages 3570–3576, Chongqing, China, 2017.

141 S. G. Loizou and K. J. Kyriakopoulos. *Automatic synthesis of multiagent motion tasks based on LTL specifications*. In *IEEE Conference on Decision and Control*, volume 1, pages 153–158, 2004.

142 T. Lozano-Perez and M. Wesley. An algorithm for planning collision-free paths among polyhedral obstacles. *Communications of the ACM*, 22(10): 560–570, 1979.

143 B.D. Luders, M. Kothari, and J.P. How. Chance Constrained RRT for Probabilistic Robustness to Environmental Uncertainty. In *AIAA Guidance, Navigation, and Control Conference (GNC)*, pages 1–21, Toronto, Canada, August 2010.

144 J. Luo, H. Ni, and M. Zhou. Control program design for automated guided vehicle systems via petri nets. *IEEE Transactions on Systems, Man, and Cybernetics: Systems*, 45(1):44–55, 2015.

145 H. Ma and S. Koenig. AI Buzzwords Explained: Multi-Agent Path Finding (MAPF). *AI Matters*, 3(3):15–19, 2017.

146 H. Ma, W. Hoenig, L. Cohen, T. Uras, H. Xu, S. Kumar, N. Ayanian, and S. Koenig. Overview: A Hierarchical Framework for Plan Generation and Execution in Multirobot Systems. *IEEE Intelligent Systems*, 32(6): 6–12, 2017.

147 C. Mahulea and M. Kloetzer. Planning mobile robots with boolean-based specifications. In *IEEE Conf. on Decision and Control*, Los Angeles, CA, USA, 2014.

148 C. Mahulea and M. Kloetzer. Robot Planning based on Boolean Specifications using Petri Net Models. *IEEE Transactions on Automatic Control*, 7(63):2218–2225, 2018.

149 M. Maimone, J. Biesiadecki, E. Tunstel, Y. Cheng, and C. Leger. Surface navigation and mobility intelligence on the Mars Exploration Rovers. In *Intelligence for Space Robotics*, pages 45–69. TSI Press, 2006.

150 A. Makhorin. *GNU Linear Programming Kit*.

151 E. Masehian and M.R. Amin-Naseri. A Voronoi diagram-visibility graph-potential field compound algorithm for robot path planning. *Journal of Robotic Systems*, 21(6):275–300, 2004.

152 M.A. Memon. *Computational logic: Linear-time vs. branching-time logics.* Graduate Course Notes, Simon Fraser University, Burnaby, Canada, April 2003.

153 O. Michel. Webots: Professional Mobile Robot Simulation. *International Journal of Advanced Robotic Systems*, 1(1):39–42, 2004.

154 D.P. Miller, I.R. Nourbakhsh, and R. Siegwart. Robots for Education. In *Handbook of Robotics*, pages 1283–1301. Springer, 2008.

155 R. Milner. *Communication and Concurrency*. Prentice-Hall, 1989.

156 M. Morari and J.H. Lee. Model Predictive Control: Past, Present and Future. *Computers & Chemical Engineering*, 23(4):667–682, 1999.

157 T. S. Motzkin, H. Raiffa, G. L. Thompson, and R. M. Thrall. The double description method. In H.W. Kuhn and A.W. Tucker, editors, *Contributions to Theory of Games*, volume 2. Princeton University Press, Princeton, NJ, 1953.

158 T. Murata. Petri nets: Properties, analysis and applications. *Proceedings of the IEEE*, 77(4):541–580, 1989.

159 N. Nilsson. A mobile automaton: an application of artificial intelligence techniques. In *Proc. Int. Joint Conf. on Artificial Intelligence (IJCAI)*, pages 509–520, Washington DC, USA, 1969.

160 N.J. Nilsson. *Problem-solving Methods in Artificial Intelligence*. McGraw-Hill, 1971.

161 J.E. Normey-Rico, J. Gómez-Ortega, and E.F. Camacho. A Smith–predictor–based Generalised Predictive Controller for Mobile Robot Path–tracking. *Control Engineering Practice*, 7(6):729–740, 1999.

162 D. Obdrzalek and A. Gottscheber, editors. *Research and Education in Robotics - EUROBOT 2011*, volume 161 of *Communications in Computer and Information Science*. Springer, Germany, 2011.

163 G. Oriolo, A. De Luca, and M. Vendittelli. WMR Control Via Dynamic Feedback Lineralization: Design, Implementation, and Experimental Validation. *IEEE Transactions on Control Systems Technology*, 10(6): 835–852, November 2002.

164 D. Panescu, M. Kloetzer, A. Burlacu, and C. Pascal. Artificial intelligence based solutions for cooperative mobile robots. *Journal of Control Engineering and Applied Informatics*, 14(1):74–82, 2012.

165 L. Parrilla, C. Mahulea, and M. Kloetzer. RMTool: recent enhancements. In *IFAC-PapersOnLine*, volume 50, pages 5824–5830, 2017.

166 T. Petrinic, E. Ivanjko, and I. Petrovic. AMORsim - A Mobile Robot Simulator for Matlab. In *Int. Workshop on Robotics in Alpe-Adria-Danube Region*, Balatonfred, Hungary, June 15–17, 2006.

167 M.-T. Pham and K.T. Seow. Discrete-event coordination design for distributed agents. *IEEE Trans. on Automation Science and Engineering*, 9(1):70–82, 2012.

168 R. Philippsen. *Motion Planning and Obstacle Avoidance for Mobile Robots in Highly Cluttered Dynamic Environments*. PhD Thesis, École Polytechnique Fédérale de Lausanne, Lausanne, Switzerland, 2004.

169 R. Philippsen and R. Siegwart. *An interpolated dynamic navigation function*, pages 3782–3789. *IEEE International Conference on Robotics and Automation*, IEEE, April 2005. Barcelona, Spain.

170 Nir Piterman, Amir Pnueli, and Yaniv Sa'ar. Synthesis of reactive (1) designs. In *International Workshop on Verification, Model Checking, and Abstract Interpretation*, pages 364–380. Springer, 2006.

171 A. Pnueli. The temporal logic of programs. In *Proc. of the 18th Annual Symposium on Foundations of Computer Science*, SFCS '77, pages 46–57, Washington, DC, USA, 1977. IEEE Computer Society. doi: 10.1109/SFCS.1977.32.

172 Quantum Signal, LLC. ANVEL Simulator. Available at http://anvelsim.com, 2014.

173 M. Quigley, K. Conley, B. Gerkey, J. Faust, T.B. Foote, J. Leibs, R. Wheeler, and A.Y. Ng. *ROS: an open-source Robot Operating System. Open-Source Software Workshop. IEEE International Conference on Robotics and Automation*, IEEE, 2009.

174 M.-M. Quottrup, T. Bak, and R. Izadi-Zamanabadi. Multi-robot motion planning: A timed automata approach. In *ICRA 2004 IEEE International Conference on Robotics and Automation*, pages 4417–4422, New Orleans, USA, April 2004.

175 M. Rahim, M. Boukala-Ioualalen, and A. Hammad. *Petri Nets Based Approach for Modular Verification of SysML Requirements on Activity Diagrams*. In *PNSE'14: Int. Workshop on Petri Nets and Software Engineering*, 2014, a satellite event of Petri Nets 2014.

176 D.R. Ramírez, T. Álamo, E.F. Camacho, and D. Muñoz de la Peña. Min–Max MPC Based on a Computationally Efficient Upper Bound of the Worst Case Cost. *Journal of Process Control*, 16(5):511–519, 2006.

177 J.B. Rawlings and D.Q. Mayne. *Model Predictive Control: Theory and Design*. Nob Hill Publishing, LLC, 2009.

178 S.A. Reveliotis and E. Roszkowska. Conflict Resolution in Free-Ranging Multivehicle Systems: A Resource Allocation Paradigm. *IEEE Trans. on Robotics*, 27(2):283–296, 2011.

179 S. Robla-Gmez, V. M. Becerra, J. R. Llata, E. Gonzlez-Sarabia, C. Torre-Ferrero, and J. Prez-Oria. Working together: A review on safe human-robot collaboration in industrial environments. *IEEE Access*, 5: 26754–26773, 2017.

180 E. Roszkowska and S. Reveliotis. A distributed protocol for motion coordination in free-range vehicular systems. *Automatica*, 49:1639–1653, 2013.

181 C. Rust and M. Gruenewald. *Petri net based design of a multi-robot scenario - a case study*. In *IEEE Conf. on Systems, Man and Cybernetics*, 2004.

182 P. Schillinger, M. Bürger, and D. Dimarogonas. Simultaneous task allocation and planning for temporal logic goals in heterogeneous multi-robot systems. *The International Journal of Robotics Research*, 37 (7):818–838, 2018.

183 T. Schlipf, T. Buechner, R. Fritz, M. Helms, and J. Koehl. Formal verification made easy. *IBM Journal of Research and Development*, 41 (4-5):567–576, 1997.

184 K.T. Seow, C. Ma, and M. Yokoo. *Multiagent planning as control synthesis*. In *Third Int. Joint Conf. on Autonomous Agents and Multiagent Systems*, pages 972–979, 2004.

185 L.G. Shapiro and G.C. Stockman. *Computer Vision*. Prentice Hall, New Jersey, 2001.

186 N. Sharma, S. Thukral, S. Aine, and P.B. Sujit. A virtual bug planning technique for 2D robot path planning. In *Annual American Control Conference*, pages 5062–5069, Milwaukee, WI, USA, 2018.

187 G. Sharon, R. Stern, M. Goldenberg, and A. Felner. The increasing cost tree search for optimal multi-agent pathfinding. *Artificial Intelligence*, 195:470–495, 2013.

188 G. Sharon, R. Stern, A. Felner, and N. Sturtevant. Conflict-based search for optimal multi-agent pathfinding. *Artificial Intelligence*, 219: 40–66, 2015.

189 T.B. Sheridan. Human robot interaction: Status and challenges. *Human Factors*, 58(4):525–532, 2016.

190 J. R. Shewchuk. General-Dimensional Constrained Delaunay and Constrained Regular Triangulations, I: Combinatorial Properties. *Discrete & Computational Geometry*, 39:580–637, 2008.

191 R. Siegwart and I. Nourbakhsh. *Introduction to Autonomous Mobile Robots*. A Bradford book. The MIT Press, USA, 1st edition, 2004. ISBN 026219502X.

192 M. Silva, E. Teruel, and J.-M. Colom. Linear algebraic and linear programming techniques for the analysis of P/T net systems. *Lecture on Petri Nets I: Basic Models*, 1491:309–373, 1998.

193 M. Song, J. Yang, Y. Wang, C. Yu, and D. Zhao. Path planning algorithm based on an improved artificial potential field for mobile service robots. In *IEEE International Conference on Intelligence and Safety for Robotics*, pages 441–445, Shenyang, China, 2018.

194 S.W. Squyres. *Roving Mars. Spirit, Opportunity, and the Exploration of the Red Planet*. Hyperion, New York, 1st edition, 2005.

195 A. Stentz. *The Focussed D* Algorithm for Real-Time Replanning*. In *Proc. of the Int. Joint Conference on Artificial Intelligence*, August 1995.

196 The MathWorks. MATLAB® 2018b. Natick, MA.

197 A. Theorin, K. Bengtsson, J. Provost, M. Lieder, C. Johnsson, T. Lundholm, and B. Lennartson. An event-driven manufacturing information system architecture for industry 4.0. *International Journal of Production Research*, 55(5):1297–1311, 2017.

198 A. Tiwari and G. Khanna. Series of abstractions for hybrid automata. In *Fifth International Workshop on Hybrid Systems: Computation and Control*, Stanford, CA, 2002.

199 F. Torrisi and M. Baotic. Matlab interface for the CDD solver. Available at http://control.ee.ethz.ch/~hybrid/cdd.php.

200 F. Tricas, F. Garcia-Valles, J-M. Colom, and J. Ezpeleta. *A Petri Net Structure-Based Deadlock Prevention Solution for Sequential Resource Allocation Systems*. In *Proc. of the IEEE Int. Conf. on Robotics and Automation*, pages 271–277, 2005.

201 J. Tumova and D. Dimarogonas. Multi-agent planning under local LTL specifications and event-based synchronization. *Automatica*, 70:239–248, 2016.

202 S. G. Tzafestas. *Introduction to Mobile Robot Control*. Elsevier, 2013.

203 M.Y. Vardi. Probabilistic linear-time model checking: An overview of the automata-theoretic approach. In *Lecture Notes in Computer Science*, volume 1601, pages 265–276. Springer, Berlin/Heidelberg, 1999.

204 M.Y. Vardi. Branching vs. linear time: Final showdown. In *Lecture Notes in Computer Science, volume 2031*, pages 1–22. Springer-Verlag, London, UK, 2001.

205 M. Velev. Efficient Translation of Boolean Formulas to CNF in Formal Verification of Micoprocessors. In *Asia and South Pacific Design Automation Conference*, pages 310–315. IEEE Press, 2004.

206 E. Vitolo, C. Mahulea, and M. Kloetzer. *A computationally efficient solution for path planning of mobile robots with boolean specifications*. In *21st International Conference on System Theory, Control and Computing (ICSTCC)*, pages 63–69, 2017.

207 E. Vitolo, C. Mahulea, and M. Kloetzer. *Path-planning in discretized environments with optimized waypoints computation*. In *23rd IEEE International Conference on Emerging Technologies and Factory Automation (ETFA)*, pages 729–735, 2018.

208 C. Wang, L. Xiaofeng, Y. Xianqiang, F. Hu, A. Jiang, and Y. Chenguang. Trajectory tracking of an omni-directional wheeled mobile robot using a model predictive control strategy. *Applied Sciences*, 8(2):231–241, 2018.

209 F.-Y. Wang. A Petri-net coordination model for an intelligent mobile robot. *IEEE Trans. on Systems, Man and Cybernetics*, 21(4):777–789, 1991.

210 X. Wang, M. Kloetzer, C. Mahulea, and M. Silva. *Collision avoidance of mobile robots by using initial time delays*. In *CDC'2015: 54th IEEE Conference on Decision and Control*, pages 324–329, 2015.

211 R. Wein, J.P. van den Berg, and D. Halperin. *The visibility-voronoi complex and its applications.* In *Annual Symposium on Computational Geometry*, pages 63–72, 2005.

212 P. Wolper. Constructing automata from temporal logic formulas: a tutorial. In *Lectures on formal methods and performance analysis: first EEF/Euro summer school on trends in computer science*, volume 2090, pages 261–277. Springer, 2001.

213 P. Wolper, M.Y. Vardi, and A.P. Sistla. Reasoning about infinite computation paths. In E. Nagel et al., editors, *IEEE Symposium on Foundations of Computer Science*, pages 185–194, Tucson, AZ, 1983.

214 T. Wongpiromsarn, U. Topcu, and R.R. Murray. *Receding horizon temporal logic planning for dynamical systems.* In *IEEE Conference on Decision and Control*, pages 5997–6004, 2009.

215 N. Wu and M. Zhou. Modeling and deadlock avoidance of automated manufacturing systems with multiple automated guided vehicles. *IEEE Trans. Systems, Man and Cybernetics*, 35(6):1193–1201, 2005.

216 N. Wu and M. Zhou. Shortest routing of bidirectional automated guided vehicles avoiding deadlock and blocking. *IEEE/ASME Transactions on Mechatronics*, 12(1):63–72, 2007.

217 N. Wu and M.C. Zhou. Modeling and deadlock control of automated guided vehicle systems. *IEEE/ASME Transactions on Mechatronics*, 9(1): 50–57, 2004.

218 G. Xue and Y. Ye. An efficient algorithm for minimizing a sum of Euclidean norms with applications. *SIAM Journal on Optimization*, 7: 1017–1036, 1997.

219 G. Yasuda. *Distributed Coordination of Multiple Robot Systems Based on Hierarchical Petri Net Models*, pages 602–613. Springer Berlin Heidelberg, Berlin, Heidelberg, 2013.

220 J.-Y. Yen. Finding the k shortest loopless paths in a network. *Management Science*, 17(11):712–716, 1971.

221 H.M. Zhang, M.L. Li, and L. Yang. Safe Path Planning of Mobile Robot Based on Improved A* Algorithm in Complex Terrains. *Algorithms*, 11 (4):44–62, 2018.

222 Y. Zhou, H. Hu, Y. Liu, and Z. Ding. Collision and deadlock avoidance in multirobot systems: A distributed approach. *IEEE Transactions on Systems, Man, and Cybernetics: Systems*, 47(7):1712–1726, 2017.

223 Y. Zhou, H. Hu, Y. Liu, S. Lin, and Z. Dingm. A distributed approach to robust control of multi-robot systems. *Automatica*, 98:1–13, 2018.

224 L. Zlajpah. Simulation in Robotics. *Mathematics and Computers in Simulation*, 79(4):879–897, 2008.

Index

Path Planning of Cooperative Mobile Robots Using Discrete Event Models, First Edition.
Cristian Mahulea, Marius Kloetzer, and Ramón González.

IEEE PRESS SERIES ON SYSTEMS SCIENCE AND ENGINEERING

Editor:
MengChu Zhou, *New Jersey Institute of Technology and Tongji University*

Co-Editors:
Han-Xiong Li, *City University of Hong-Kong*
Margot Weijnen, *Delft University of Technology*

The focus of this series is to introduce the advances in theory and applications of systems science and engineering to industrial practitioners, researchers, and students. This series seeks to foster system-of-systems multidisciplinary theory and tools to satisfy the needs of the industrial and academic areas to model, analyze, design, optimize and operate increasingly complex man-made systems ranging from control systems, computer systems, discrete event systems, information systems, networked systems, production systems, robotic systems, service systems, and transportation systems to Internet, sensor networks, smart grid, social network, sustainable infrastructure, and systems biology.

1. *Reinforcement and Systemic Machine Learning for Decision Making*
   Parag Kulkarni
2. *Remote Sensing and Actuation Using Unmanned Vehicles*
   Haiyang Chao and YangQuan Chen
3. *Hybrid Control and Motion Planning of Dynamical Legged Locomotion*
   Nasser Sadati, Guy A. Dumont, Kaveh Akbari Hamed, and William A. Gruver
4. *Modern Machine Learning: Techniques and Their Applications in Cartoon Animation Research*
   Jun Yu and Dachen Tao
5. *Design of Business and Scientific Workflows: A Web Service-Oriented Approach*
   Wei Tan and MengChu Zhou
6. *Operator-based Nonlinear Control Systems: Design and Applications*
   Mingcong Deng
7. *System Design and Control Integration for Advanced Manufacturing*
   Han-Xiong Li and XinJiang Lu
8. *Sustainable Solid Waste Management: A Systems Engineering Approach*
   Ni-Bin Chang and Ana Pires

9. *Contemporary Issues in Systems Science and Engineering*
Mengchu Zhou, Han-Xiong Li, and Margot Weijnen
10. *Decentralized Coverage Control Problems For Mobile Robotic Sensor and Actuator Networks*
Andrey V. Savkin, Teddy M. Cheng, Zhiyu Li, Faizan Javed, Alexey S. Matveev, and Hung Nguyen
11. *Advances in Battery Manufacturing, Service, and Management Systems*
Jingshan Li, Shiyu Zhou, and Yehui Han
12. *Automated Transit: Planning, Operation, and Applications*
Rongfang Liu
13. *Robust Adaptive Dynamic Programming*
Yu Jiang and Zhong-Ping Jiang
14. *Modeling and Control of Uncertain Nonlinear Systems with Fuzzy Equations and Z-Number*
Wen Yu and Raheleh Jafari
15. *Path Planning of Cooperative Mobile Robots Using Discrete Event Models*
Cristian Mahulea, Marius Kloetzer, and Ramón González